DEEP ECOLOGY

EDITED BY MICHAEL TOBIAS

Published in 1985 by Avant Books

3719 Sixth Avenue
San Diego, California 92103

Library of Congress Cataloging in Publication Data

Deep ecology.

1. Ecology – Addresses, essays, lectures. I. Tobias, Michael.
QH541.145.D44 1984 574.5 84-9176
ISBN 0-932238-13-0

Produced by The Word Shop, San Diego
Cover art and book design by Ed Roxburgh

Printed in the United States of America
First Printing

Table of Contents

ILLUSTRATIONS/PHOTOGRAPHS

Introduction
Michael Tobias

There was never a time when human beings did not appraise the natural world, painfully aware of their own paradoxical position within it. Paleolithic artisans recorded infinite details of death and resurrection in the hallowed caves of Europe and the Near East, evidencing a strong knowledge of plant and animal species as well as the fact of their own separation from them. By their own evolved consciousness, hunters of the late Pleistocene epoch were able to exploit the surrounding ecosystems. But the price they paid was psychological alienation. The subtlety of these many paintings strikes an ambivalent chord of angst and wonderment. The inherent contradiction – passionate identification with nature, and the simultaneous exploitation of natural resources – is at the heart of the deep ecology issue.

This duality was sustained in later Egyptian, Greek, and Roman times. Osiris, Apollo, and Cybele were deities of nature who were tempered with all the eccentricities and inconstancy of human nature. Plato looks out on the benevolent shores of his homeland and deciphers crisis in the guise of topsoil erosion. Lucretius laments the insensitivity towards the rural farmlife of his technocratic contemporaries. During the Renaissance, Paradise is perceived somewhere in the dim past, whereas the present is full of apocalyptic premonition. By the 18th century, the Gods had left Nature to mankind. George Washington worried about that, noting the excessive rage of land speculators in post-revolutionary America. The American Revolution was largely financed by the first wave of real estate moguls. Karl Marx impugned these spurious relations upheld by capitalism, noting that the exploitation of labor was the exploitation of nature. The Vermont naturalist and farmer George Harsh furthered this analysis in his book *The Earth as Modified by Human Action* (1874). Marsh perceived the eventual downfall of America as a result of the American public's loss of proper relation to the Earth.

In the 20th century this ecological schism has shadowed the entire planet, pitting the original insights of our forebears against the mechanistic priorities of latter day comfort and ego. It has been the habit of subjectivism in the arts and – more recently – the natural sciences, to oppose these two historical tendencies, drawing upon the devout poetry of the one, and the insanity of the other. At the turn of the century, during the vogue for works like *Call of the Wild,* conservationism arose as the ethical means of redressing human nature gone awry. From its inception, the preservation of Yellowstone, the Adirondacks, and Yosemite there were deeply enflamed political and economic motives exerting persuasive cause and effect. The outcome of the earliest congressional debates suggests a tenor to environmentalism which was, and which continues to be, pragmatic, remedial, reactive, and short-term in focus. The abiding principle was not the value of nature in its own right, but the value of nature for man. Alexis De Tocqueville first pointed to the American penchant in this regard, remarking with some disdain on the arrogance of power at work in the forests and out along the prairies. When painter/ethnographer George Catlin cried out for the preservation of the Indians, it was as if a museum curator were endeavoring to protect his most highly prized specimens.

This patronization has taken an especially troubling turn during this century. Conservationism has not addressed such nebulous matters as 'the love of nature,' but rather has sought to protect mankind's resources, his god-given oil shale, abundant timber, and teeming continental shelves. The Law of the Sea conferences and Antarctic Treaty debates clearly indicate civilization's intention of leaving no part of the planet unexploited.

Environmentalism has had to confront a shift in logic not simply semantic: What John Muir took to be spiritual inviolability, later conservationists interpreted to mean harvestable. Yosemite would be "democratized," roads built, and all the conveniences of urban life engendered there. In essence, Yosemite would become a major tourist resource for America. The meaning of conservationism was thus, from the very outset, a compromise position.

It was spelled out at the first National Governors' Conference convened by Teddy Roosevelt in 1908, an august affair that sought to examine America's stewardship of natural resources. The United States had shortly before adopted a diplomatic measure granting this nation the god-given right to use chemical warfare if it damn well pleased. Roosevelt himself had only recently returned from one of his many hunting forays, this time from Asia, where he managed to bag a Giant Panda bear. It might seem unfair to assert such condemnations, given the climate of those times, but that is the point. The early 20th century context for addressing ecological matters has not changed appreciably. If anything, with the emergence of fission, and the multinational corporation, the shallowness of our conservationist impulses has been greatly augmented.

Other conventions met in the wake of Roosevelt's, all of them oriented to a doctrine of preserving nature specifically for human utilization. This shallow approach to the natural world has its counterpart in the "deep" ecological outlook, a legacy of moral and aesthetic concerns beyond mere utilitarianism.

Deep Ecology concerns those personal moods, values, aesthetic and philosophical convictions which serve no necessarily utilitarian, nor rational end. By definition their sole justification rests upon the goodness, balance, truth and beauty of the natural world, and of a human being's biological and psychological need to be fully integrated within it. This is a premise easily ignored in our world, where the possessive case – our – increases unchecked with each new bit of legislation, industrial outreach and scientific hubris. This technological and conceptual tendency of civilization has served to isolate our species from all others, save perhaps for certain rodents, insect scavengers, French poodles, Siamese cats, horses, spider plants, and marigolds.

This isolation, or lack of wilderness, is biologically bereft of the very connections which normally assure a vital place in the planetary puzzle. The loons have been on Earth for 130 million years. Whales and wolves a little less. Human beings, a mere 2 million years. That we are in a process of current speciation may be suggested by the fact we have doubled our lifespan in the past 10,000 years. We might well discover that the Parisian of the year 2000 would be incapable of procreating with a representative of the Cro-Magnon.

Cybernetics and the evolution of computers adds yet another twist to human destiny. This is a water planet. The first terrestrial matter was probably silicon, beaches. From a computer's perspective, human beings are on Earth merely to reorganize silicon into conductive chips. From the biosphere's perspective, the whole point of Homo sapiens is their armpits, aswarm with 24.1 billion bacteria. The purpose of the brain's evolution is strictly for transport of armpits to diversified niches, perhaps other planets. For the Earth stares out into space and is alone, aware of itself, yearning to share the gift of life. Ironically, we are the one species that can focus on the stars and for whom the stars have meant something fantastic. The biosphere has happily managed to articulate through us and yet we tempt her with catastrophe. This paradox is also at the heart of deep ecology.

The assorted reflections, essays, poetry, and dialogue here presented come in the wake of considerable previous labors of love – books, conferences on ecological humanism, and at least four millennia of diverse revelations and insights. Three wildly separated examples of what I mean come to mind. First, "The Dispute of a Man with His Ba," written 2200 years ago by Khety during the Twelfth Dynasty under Amenemhet. The Ba refers to the seeker's soul. With his face into the breeze, this man sits "longing for home, the clearing of the sky, a rain-washed path and the smell of lotuses." He is an existentialist hero who finds meaning to life in simple attributes of nature. In Burton Watson's sublime translations of the T'ang poet Han-shan, one glimpses the radical hermit and his powerful faith in nature:

> *I sit leaning against the cliff while the*
> *years go by,*
> *Till the green grass grows between my feet*
> *And the red dust settles on my head,*
> *And the men of the world, thinking me dead,*
> *Come with offerings of wine and fruit to lay*
> *by my corpse.*
> [*Cold Mountain* – (New York: Columbia University Press, 1970), p. 114].

Finally, there is Thoreau writing, "No idea is so soaring but [that] it will readily put forth roots; No thought but is connected as strictly as a flower, with the Earth. The mind flashes not so far on one side, but its rootlets, its spongelets, find their way instantly to the other side into a moist darkness, uterine."

Deep Ecology presents a way into the study of ecological values. Merging fiction and nonfiction, history, science, art, philosophy, and literature, its emphasis quite clearly rests upon an interdisciplinary, humanistic approach to questions of unending relevancy.

The three sections of this book encompass a broad cross-section, suggestive of the movement's multitude of concerns. "Hard Numbers" pertains to the analytical, history, facts, trends; the two edged sword of technology; the reappraisal of means and ends; the economics and politics of transformation.

"Heartland" emphasizes the "right-brain" position. Visions, totemic symbolism, the feelings and senses. This experiential agenda complements a rationalist's world view. Humanism is a word etymologically indebted to humus, and humility. Similarly, the ancient Hebrew word for Earth was *adamha*, referring as well to Adam, the first man. The word nature comes from *nasci*, the feminine word meaning to be born. All of these connections are at the root of our collective unconscious. This second section of the book explores them.

"Awareness and Reason" proposes several philosophical frameworks, tools by which to integrate thought and action, desire, and design. "A philosophy has to be a synthesis of theory and practice," Arne Naess reminds us.

We hope that *Deep Ecology* spawns increasing interest in this direction. The book is by no means comprehensive. That would be a difficult task, considering the indisputable breadth of such concerns, underlying all of human existence.

SECTION ONE

HARD NUMBERS

Humanity and Radical Will: Reflections from the Island of Life

Michael Tobias

Formerly an Assistant Professor of Environmental Affairs and the Humanities at Dartmouth College, with a Ph.D. from the University of California-Santa Cruz, Dr. Tobias has written and edited several books, including The Mountain Spirit *(Viking, 1979),* Deva *(Avant Books, 1982) and* The Mountain People *(Interprint, New Delhi). He has written feature films for Dino De Laurentiis Film Corporation, and written and produced programming for the networks.*

THE SEARCH

All of our spiritual myths are recent ones, imbued with the technological metaphor. The Egyptians of the Old Kingdom worshipped the dung beetle, Scarabaeus Sacer; the Akam-speaking West Africans revered the Kwaska Anase spider. The Sisyphean coprophagist (shit-eater) beetle rolled its sparkling take of feces with a celebrated mechanical dexterity later emulated in the building of the pyramids at Saqqarah and Gizah. The spider, in turn, taught exemplary weaving techniques. The Bible is governed by its own set of mechanics – principles of agriculture, architecture, strict kinship, violent taboos, genocidal warfare, and a rags-to-riches theory of economics. Its emphasis on monotheism in Deutero-Isaiah (later borrowed by Zarathrustra, Christ, and Mohammad) marks a persuasive advance in *rampage logic*. What is this singular

God, this power in the universe, about which knowledge and fear may be instilled by simply reciting the Shema, "Hear, Oh Israel!"? What are these commandments, oral histories and prophetic harangues; the wide swathes of warrior demolition and ego; enumeration of friend and foe, warring families, irate nation states? The coercive elements of a battle style were forward in their thinking, alliterative, given to broad sweep, solicitous detail, employment, elegant, apocalyptic nerve. From stigmata to keeloid nuclear scars, the religious conviction obtains through carved totems, each face gouged out, carefully crafted until a wrathful shrill emerges in the snarled nostrils and eagle eyes; until the first autocrat squelches that small voice of life.

The metaphor may have gone in an opposite direction, towards peace. But it did not. Rather it assumed the fierce unyielding ego projection of scapegoats, fear of the future. Such bias evolved out of suffering (of that we can be clear), for there has never been such an ongoing, 7,000-year backlash of getting even with God for making us the way we are, as a child blames its mother. The Bible strains terrifically to acquire management, monoculture, to delegate assignments to a living coterie of ranters and ravers. It is modernist in its appeal. It tries to outsmart, gun down, and reinvent the older verities. What forces created the Bible's particularly painful presumptions? What bleached bones, collapse of ecosystems, and psychic moonlight accomplished so full a turnabout?

The Bible suggests a final frantic effort to wrest control from earlier totemic, multiple-thinking, animistic forces; from these tribes legions still retaining the cultural context of the surrounding cornucopia. Imagination is the metaphor we're referring to; the synthesis by which our neural excess bridges over smooth flowing waters all around us. We build bridges out of danger.

Today the two approaches stand in tragic contrast. The last remaining tributes to original consciousness are embodied in those tribal groups whose life is synonymous, more or less, with an approach to original systems, linked in so many fruitful ways to the essential activities of earth and sea in concert with atmospheres. Perception of such frenzy — animal migrations, seasonal growth increments, lightning upon forest, rain upon meadow, tides across reef — must have analogies in the brain, restless areas ripe for the environmental imprint and able, with enough exposure, to formulate such events over again in the abstract and apply their predictions to human behavior. Modern, derivative mind on the other hand, is explaining away the continental shelf, the beach cliffs from which we are drifting, sanctifying with the aid of science our encrippled distance from nature. This modern derived mind justified its first manipulation of other human beings, and in their name, of all other species, of all matter. This distinguishes the curious, primitive, *original traits* from the imaginative, *derived ones*. We see that aboriginal groups today remain essentially undifferentiated, stable, endowed with the same set of physical characteristics, a similar mental apparatus, and precisely the same relationship to metaphor as that which pervaded their earlier cousins. The function of symbolism is the same, the size of population, hegemony, distribution of resources, and repertoire of survival techniques— all similar. Primitive man's brain has accommodated itself to pre-existing conditions, because they were *his* conditions. The route of his evolution has been checked at the locus of efficiency and well-being. On the evidence of the past million years, it seems reasonable to assert his success. If he has no 747s, no metal zippers for closing up his fly, it is not for lack of inventiveness.

The derived mind can forcefully draw interesting conclusions about most things—from DNA, chemical warfare among insects, deep-sea vents, the bacteria in snow crystals atop Mount Everest, the history of Portugal, the invention of the lever, to Barnard's star, and cultivation of the kiwi fruit. But how has this range of reality helped us?

Since the beginning of the Sixteenth century, when Sir Thomas More outlawed all warfare in his Latin treatise *Utopia, Homo sapiens* have fought 250,000 major battles, armed with explosives. Complexity, it would appear, has polarized differences, not softened them. The Pygmies and Hopi are certainly not without complexity. What is it, then, which works to mollify and smooth over all differences in those two societies? One of the very first distinctly human traits must have been the desire to acquire, and in so acquiring, to become something exquisite and novel, reborn, redefined. The lover desired to become his mate; the transfixed beholder of nature sought to become the thing in itself, and this argument triggered what we know to be the history of art and of myth. Such desire is fixed at the first moment of goodness, the flexibility to restrain ourselves for the idea of paradise regained. The infant clings to the breast.

Such capacity was perverted into the greed of regaining. While the *desire* itself remained, its self-importance would covet and repulse, steal and kill to get at its vision, horde its gold, its chattel. Such perpetuation of envy ultimately lost sight of pure desire, the desire that brought man's evolutionary intelligence to bear with the aid of nature. We became unnatural. We recognized no world but the world within ourselves.

Julian Jaynes has remarked that schizophrenia was pervasive 3,000 years ago. Feeling and information fought fierce battles, came out eruptively in the visitation of voices, and the edifice of hieroglyphic incantations. It was the transitional stage on the eve of a monstrous physiological occurrence: merging of the neocortex. Two-hundred-million fibers of the corpus callisum came together. The two lizards did not give up their wrath. They were merely repressed.

It is repressed consciousness which most clearly defines thought from that time; anger aggravated by a guilt-ridden devotion to imagination's fondest comforts, comforts dumbly derived, unproven, unworked for, without reason enough, at any rate, to justify the imbalances of acquisition. This is an age of the selfish mind. And for all of the good it has wrought (though who can judge?) our unstoppable imagination barrels ahead in the night like a train with its headlights turned inward (Pasternak's simile). The anticipation of a crash — spectacular, like the day of His wrath pouring in — stains the otherwise unsullied faith of everyone alive, prepares the bed of roses, lets us know that good and evil are going to smash heads, no escaping the conflict.

The modern search for reason is like a surgery of switchblades, to be in touch not just for the duration of an idyllic paleolithic swoon, a fleeting afternoon, but to be so incorrigibly, intimate with the whole of life, beyond this blinded generation. What rebel of thought remains to recapture the holy primitive? What fundamental font, able to go left, to go right, still breathes and pulsates; what starlight, what deep source of integrity may yet move to temperate office?

Solid ground on risen rules, euphemistic extremes, frosted platforms of performance from which poetry, unlike science, need, can go no further. These peaks, this glacier cirque have been lifted into symbolic identity in our minds over eons. No human fledgling, for all of his intellection, will add to them. If anything, by the power of curiosity he is liable to invade their pristines, debate jurisdiction, strip them of timber, tap hydroenergy, pollute the water, overcrowd the slopes. Like a dream of lovers, the private intercourse of bronze nudes, fingers straining and adrift on Sistine Chapels, poetics — like the mountains — are tautological, an end in themselves, whose only possible history — like the history of ice water — is a chain of sighs and inward looking stares, the bright, stained glass and quiet felicities they might inspire among some of those who come afterward with eyes opened. No jet airplanes or wonder drugs will result of such poetry and mysticism. True enough, since I refer to it, northern Thai or Vietnamese montagnards, upland Taoists, and Andean shamans have cultivated the mysticism of science from their mountains, a proven trove of herbs and millenialist cures; have passed down a fabulous inheritance of remedial and playful wisdom; a bedazzling array of explanations for everything. Where there is local custom, there will be local wisdom. The human heart — like poetry — will always reside locally. But science, as characterized by the annals of material progress, requires a wholly enlarged space in which to operate — some other sequence of events, where ideas build upon ideas, a scaffolding rich and famous, like the technology that puts the formulas to work. Poetry may exploit a theme, a simile, an obsession, even an image or two; but its personality is more feverishly stamped with immutables that have a bottom line in the eccentricity of their maker; are more rudimentary than the perennial purpose of science. The mind, arguably, may catapult evolution, or destroy the planet, or entrap, fall into eventual accord with divinities at work in and around us – nature acknowledging itself, as Marx voiced. When science delves deeply and lonesomely into nature for a solution, it begins to look like poetry, its formulations hailed as "elegant." These two spectra, poetry and science, are addled flirts. Between them is the human experience.

Their equal store of ignition power will never be satisfied, or at least we do not yet evidence it. Clashing armies, dynasties and egos, ideas supplanting other ones, pulp, is the motion of the mind's fancy as it eyes toward innumerable horizon lines and futures, reconcatenates the self, charges being with an invigorated outside law, a synthesis of unknown quantity that defies ordinary arithmetic, so that two plus two equals paradise. Our one and only hope. We hope for more than ourselves, in other words. Clumsily, we search for nature. The rainbow, first declared a covenant in Genesis (9: 12–17) was recreated in the British nineteenth-century argot of painting. But Newton's earlier prism, said Keats, had obliterated the poetry of the rainbow. However inadequately, the attainment of Nature – if only a mere figment of thought celebrated along with lesser intoxicants – our own human nature does not come easy; it is not just inside waiting, but demands grand effort, coaxing, understudy, self-improvement. It has never been enough to merely feel deeply, to love, to be good and integral. This is the stuff of eternal worry. That something always seems to be lacking, a deferral at hand, some other advantage calculated in our secret inners. Thus, Orwell said at the opening of his essay on Gandhi, that the saint is always guilty until proven innocent. We want, we lunge after more and more. Even Buddha wanted, desperately, to know, to reconcile human heartiness with evil all around. Under the programs of science we have made fashionable such claims, have lent supreme purchase and narcissism to our genius. Whereas there has been no particular corresponding apotheosis within the poetries of experience, but just the opposite, a giving away of ourselves. Poets have had to delve into anarchy and selfishness, only to the extent of personalization, where the light of a Sinhalese monastery shines for the poet alone; and where the boil of sardine scintillates in a lonely vantage applicable to that singular eye that has recognized it. Poetica is elitist. So is sexuality, and for that matter all of biology, the stars, the planets. The Earth is an operant whole, its parts magnificently and unselfishly involved. But there is a radically different perspective, whether it be possessiveness, the aloneness of death, or the infinite shades of predation. This latter vantage, if seen in entirety from protoplasm to man, could of itself destroy our uprightness. It is consciousness that keeps us firm, holding steadfast against the collective. Poetry repudiates much of the world. Ironically, it is only through such denial, such self-absorption, that we can come full circle and sense the greater sphere outside of ourselves.

In old age, if we are lucky, our minds glean something of the completion, the poetry. Experience should help us along in this. So should science. I sit acrest a glacial erratic. Granite. Not a sound in the world other than my coursing blood. Silently, gingerly, these 14,000-foot-high aerial faces cast sensual shadows over the thinly crevassed contours. Lush shadows, slowly coming toward me in tropical plume formation, delicate, ever so curvacious. I would say, based on such evidence, that even this almighty mountain can love. To admit it, finally, is to raise a painful insight into the conundrum of mountain climbing, why we do it in so many ways. Painful because it is only the heart that perennially wails, which needs to love, would travel to the ends and heights of Earth just to let it cry out, over an otherwise intolerable abyss. This love is, by another name, paradise. What is extraordinary to me is the fact that paradise – the wilderness – has always been portrayed as somehow predating or eluding our own auspices. Intellectually, scientifically, this is true. But poetically, this is not true. We love, we *paradise* every day of our existence.

This is where the mind, which conspires with the heart, can help us to read rightly the maps, the weather, all the emotional prefigurements of classical dichotomy. The sciences give us food and shelter. The heart gives inclination. Nature provides the beginning and end of all substance, out of which we are formed, wrestle, hope.

There is a sound above me, some bird I do not recognize. Its wings bristle against the later afternoon air, pumping and displacing chilled molecules. It's going home, I imagine, down, down into treeline. I can't see it, especially not without my glasses; but I can well enough visualize its cozy niche, utterly jake, abounding in dried turf and regurgitated foodstuff – insect gruels, tuber casserole. How akin are we, though never a word transpires between us. In the interstices is a god who combines everything. And if we are intent upon discovery – what may be our potential, why are we here, where are we headed – then we will obviously need to acknowledge that bird overhead in every respect. The Chinese poets warned that until we learn the language of sea gulls we will never learn the language of ourselves. St. Francis had a similar insight, eight hundred years ago. Rachel Carson revealed our xenophobia about birds. Though Americans continue – by powers more ancient than DDT – to rally about the eagle. There are obvious reasons why we love the eagle. Other, inobvious and archetypal explanations hover all around us, like angel wings. The wings, I gather, are our own, if we will only exercise them.

BETWEEN ELEGY AND REBIRTH

Today there is sufficiency both grand and gratifying, sun glints, Earth breeze, electron frenzy, the rumble of primordial inertia; a spider's silvery strand, a linear gurgle of the creekbed, October's fair chill that soaks the frost, and tingling air, each stalk of trembling life – a world vision entire. Verities abound; in flickering tails, scuttling whisps, soil that teems with roots cross-jumbled; ruby-haloed insects the size of freckles mingling with a purposeful agenda deep beneath the sight that humans ordinarily glimpse of landscape. This seasonal agility, heat and cold, attests to marvels beyond our ken, the black and white delineations of how we'll go about our narrow life; what hope, which god, what natural selection binds us to our comforts? From where we stand, look out, and muse, there is a blushing mystery, colored by the very physics of light, the chemistry of collision, neurology of mind, the ecology of everywhere. Glacier fleas, fond of the frozen state, will die conversely by the warmth of a human palm; ladybird beetles have an innate urge to fly to the highest mountain in spring for mating, and have been found on snow patches in the western Himalayas in clusters 200,000 thick per square meter; snow buttercups and mountain aven flower under ice. The perception of variety and contrast is a blessing to be delighted in. Too often, however, our species has exchanged this gift of sentience for the promise of exclusion, the logic of opposition. Our view of the world is largely contained within the idea of it, our approach to interaction.

Life, in all its myriad interdependencies, may actually regulate and fashion the environment. This is what James Lovelock calls the "Gaia" principle. Posidinus, Aeschylus, the Ionian philosophers were aware of it, as were Middle Kingdom Egyptians, who voiced their own eloquent understanding of it in the fourteenth century B.C. *Hymn to Aten*. Now, modern science accedes to the notion: that the biosphere engenders a spectrum of regulatory mechanisms propitious to its survival, forging energy-consuming syntheses of carbohydrates, and energy-releasing assimilations of carbohydrates by other organisms, in the process determining global temperature, landforms, and majestically delicate balances. "The only feasible explanation of the Earth's highly improbable atmosphere was that it was being manipulated on a day-to-day basis from the surface, and that the manipulator was life itself," writes Lovelock. The primeval rain of archaebacteria upon cooling oceans, outgasses, protocellular redundancy, and the chemical experiments – among hundreds of billions of combinations – resulted in a simple organic molecule, eventual amino acids, cytochrome *c*, polypeptides. This was our earliest energy brokerage, our genetic code. Utilizing hydrogen from the rocks and newly born waters, bacteria manufactured their own ephemeral foodstuffs, and a major biochemical transition was at work, two billion years ago, as the planet woke up to its oxidative metabolism. With 10^{80} possible chromosomal pairs, the great mitotic square dance flung its course across the globe. Life could not get enough sex once the idea caught on. This was the eukaryotic hereditary mechanism, the life spice of algae, flatworms, ciliates, metazoa, all grappling for their moment in the shielded sun. *Cooksonia caledonica*, club mosses, ferns, flowering angiosperms—a vascular aesthetics, glory, a yearning perfection.

Then something happened: It had happened before, incurring a collapse of the food pyramids, and the massive demise of innumerable herbivore species. The meek burrowed in for the long winter. Lost amid a welter of Cenozoic disruptions, is the lonely tale of small quadropeds, able to see in three dimensions, crazily flippant with new opposable thumbs. These little fellows developed homeothermic means (fur, milk glands, larger brains) enabling them to compete outright with reptiles in an evolutionary scramble at the end of the Cretaceous era. Out of this stunning chaos hopped our true Adam, a shrew-like insectivore who saw its rainbow in the arboreal heights, prevailed upon the new aeries with subtler discriminations of intellect, and scurried around all day feeding itself. Until, within its ranks, there were constraints. The Suborder *Anthropoidea* branched out, fueled by new protein sources of meat. Chary, but ever curious, some of the little guys descended, leaped back up into the trees, and came down again over millions of years. Then they were ready to move out, and the plains of heaven bade them do so. This time they didn't look back. From that time on, their story is more familiar. Bipedal and socially organized, *Ramapithecus* fanned out from Kenya to China. Comparative osteologies suggest they favored the banks of permanent streams. Their greatest resource was their thumb, and the thumb's pre-eminent muscle, the flexor pollicis longus. The thumb afforded us precision grip, freed us to exploit, and endowed the hominids with remarkable sensitivity. That the thumb was revered in Neanderthal times is illustrated by some 200 imprints of hands with missing thumbs, on the walls of the Tiberan cave in northwestern France. To this day, when the arm reaches out towards an object, the sweat glands of the thumb are activated, a survival strategem left over from arboreal times when such sweat served to solidify our grip on bark.

CRO-MAGNON CONSCIENCE

They suffered angst, contemplated the same universe that Heraclitus, Hipparchus, Eratosthenes, would come to love. Not without his form of brutality, he drove animals to extinction, then pondered the results, attempting with crisis colors to exert shamanic control over animal mating and migration corridors. The inner penetralia of Lascaux suggest prayer, the hope to tie resources to a stable place. He found that burned over areas produced new growth of succulent grass which in turn attracted game. By 18,000 B.C. he was domesticating wheat in Egypt, and cattle. His mind was a wilderness that all future souls would yearn for. Among the Dal race in Dalarna, Sweden, the Guanches on the Canary Islands; in !Kung, Suya, or Bimin-Kuskusmin society in Papua, New Guinea, we see final vestiges of the cynegetic (hunter–gatherer) mode, and it is splendidly efficacious, quiet, modest.

THE GREAT SEVERENCE

The young domesticator was troubled. Deprived of Ge, the earth mother, and of his human tit (the mother now rearing up to sixteen children in her lifetime), he had to clutch his few possessions for whatever they could give to him. Animal symbolism was distorted in his fever, wildness bred out. The great metaphors, by which taxonomy guided the earliest polity, were abandoned. The dreamer of myths was now lost to a purgatory of debt: the future. At Jarmal, Catal Huyuk, Jerico, throughout the Tigris and Euphrates and Nile valleys, at the Spirit Cave in Northern Thailand, he was frantically consolidating for surplus. Barley at Ali Kosh, pigs at Jarmo, cattle in Greece, beans throughout Central America, by 7,000 B.C.

HOMO FABER

At the city state of Ur, 3,000 animals were slaughtered annually, allocated to temple personnel. The Mesopotamian onager (wild ass) was harnessed, then replaced by the horse, which had been steadily domesticated by pastoral nomads in the Ukraine, broken. In Sumer, a lever device gained some currency, as did clay maps; while in Upper Egypt, at al-Badari, were wheeled vehicles, stone walls, extensive irrigation. Animals were buried on the edge of villages, wrapped in linen.

THE SQUARE-ROOT

The balance was in the bias. By the time of Prometheus, the gods themselves were unsure what they were doing. So Zeus married Hera. Archimedes, in 220 B.C. formulated the first concept of limits, which culminated a mighty tradition of elegant, Ionian naturalism, of which Homer and Hypocrates spoke in aboriginal dialects, with rhyme still smelling of Shanidar. *The Politics, Ethics, Critias* all put faith in balance, interdependency; and in his nudes, Praxiteles furthered the legacy. Like the Greek stewards, Lucretious later would espouse the abnegation of technology. The Greeks and Romans had the capacity, the knowhow, to develop a mechanistic universe. They deferred this proliferation, as the marine mammals had done. The fin over the thumb.

AGRICULTURAL REVOLUTION

While the Tang Chinese were cultivating poetic naturalisms, sixth-century Taoist farmers were rotating three fields in two years for wheat and millet. They knew their land, but did not perceive the dangers of upper-basin watershed despoliation. In Fukien, following floods in 1012 A.D. was massive drought. A drought-resistant rice strain was introduced from Champa, capable of maturing in 100, rather than 150 days. It saved lives and encouraged future strain-mixing experimentation. About the same time, the Chinese invented gunpowder, used successfully at the Battle of Crecy in 1346 A.D. The Sung Dynasty landscapists (Ma Yuan, Kuo Hsi, Li Chen, etc.) had engendered resplendent, shan-shui foregrounds, with only a monastic fragment – amidst plunging cataracts and astonished deer – to indicate the human presence. But Western painters changed all that. Altdorfer's "Battle of Alexander" used a wild backdrop to cast the glowing altercation of armies plowing into one another with longbow. We see confusion in Bruegel's deep, demonic fear of technology, in his love of peasant life. Within a few decades of his "Icarus," bituminous coal smelting, Napier's adding machine, Savary's engine, the discovery of microscopic organisms, and a dozen other innovations usher in the first era of techno-fix. Columbus' *Third Letter*, Bacon's *New Atlantis*, dualistic founderings in the New World, and the emergence of the Noble Savage further exacerbated a profound western nostalgia for the past and new expectations for seemingly unlimited capital. The Puritans, equating such wealth with divine right, fostered a double-bind, which the Jeffersonians, considered "self-evident." By 1765 there was a steam engine. At the height of Romanticism, in 1791, a gas engine was sputtering its way toward immortality. "Nature is purely a matter of utility," Marx voiced. Engels, citing the work of George Marsh in Italy, quietly contradicted him. But by then, the great gulf was at work in American politics. George Catlin's poignant observations of the "natural man" on the Great Plains merely told us how far we had come from our beginnings.

HUMUS VERSUS HUBRIS

Peruvian guano, Chilean nitrate of soda, Indian saltpeter – these were early fertilizers. Even the Romans were dimly aware of nitrogen fixing. In 1837 Sir John Lawes experimented with manures and went on to patent a process for treating phosphate rock to manufacture superphosphate. Fifty years later, nitrogen-fixing bacteria were discovered. By 1913, the controversial Fritz Haber had developed a process for ammonia production. There was no stopping *Rhizobium* by that time – the breakthrough of symbiotic bacteria on legume roots. It seemed like the sky was the limit for agriproduce. Then, in 1963, Kellogg Engineering developed a process for synthesizing atmospheric nitrogen into chemical fertilizer. There was *no* limit. What with Donald Jones' earlier "double-cross" hybrid and Paul Muller's realization of the insecticidal properties of DDT, the earth had been harnessed. There were only three fallacies to this new scientific optimism: the rise of vulnerable monocultures, a postwar baby boom, and the alarming decrease in soil fertility. The womb stole from the soil. Population in Europe had grown at a rate of less than 1 percent from 600 to 1600 A.D. From 1750 to 1850, the population doubled. The impact of those years is neatly catalogued in lead deposition along Greenland's lonely icefall.

HOMO COLOSSUS

Each American's food needs consume one acre annually. But economic exponentialism does not work so humbly. Malthus saw this in the first edition of his *Essay on the Principle of Population,* bluntly, naively refuted by Adam Smith's "invisible hand." Today, more than 90 percent of the energy used by *Homo sapiens* derives from sources other than each current year's crop of vegetation or interest. This is due to a seldom-recognized caste system. The real first estate comprises a greedy cadre of "ghost slaves." There are nearly eighty of them for each woman, child, and man in the United States. Even during the height of the Depression, the Chicago World's Fair could stage its Century Of Progress without blinking an eye.

GROUND TRUTH

It is not sufficient to merely indicate that some 65 percent of all clothing in the world is now made of synthetics; that 100,000 new vehicles come off the assembly line daily. Or that the whale celebrants, apparently desirous of speaking with us, become victim to our genocidal mindset, as do millions of other species and spirits. Landsat and the Thematic Mapper yield images of a bludgeoned earth. Six-thousand square miles of forest are devoured each year – 70 acres every Sunday for the New York Times alone; two billion tons of topsoil are eroded annually, resulting in floods that killed over one million people throughout the Himalayas in the decade of the 1970s. One hundred kilometers of new desert has arisen in the Sahel in the last 30 years, while water withdrawal increases 300 percent worldwide. Arizona is shrinking; Brazil's tropical, moist forests are being ravaged by slash burners and the Correntao clearing techniques, which break the natural selection cycles, extinguish refugia, and foster regions like the Bragantina – a dead earth.

Each year 70 million additional people need housing. In one century, Americans have become fully urbanized. And in that transition, 125 million animals, inhabiting every soil-laden acre of land, are thrown into chaos, decimated; as urban conurbations exert outgrowth pressure, reducing prime agricultural lands and wild lands, to slave lands.

Despite this show of stupid, indefatigable growth, there is the inescapable Faustian bargainism. Worldwide production of wood, wool, mutton, and marine catches is declining, in some cases prodigiously. Since the mid-1970s, double-digit inflation, unemployment, and labor dissatisfaction has paralleled the global deterioration of all biological systems. Precious minerals have been exhausted, new oil discoveries severely limited, and agricultural and water resources depleted. Production costs in every sector have soared, borne aloft by factors strictly ecological, but only dimly understood as such. In the meantime, the gulf between western consumption and Third World sustenance grows by a ratio of inordinate magnitudes. The Green Revolution has masked myth, myth that serves us no less sanely or importantly than bacteria in our gut.

The "structuralist fallacy" (e.g. Aswan High Dam) preserves the master-builder ethos of excess engineering. Most children are more apt to build sand castles higher and larger than will work, no doubt for the pleasure of seeing them topple. Only in sophisticated instances does the creative pleasure of fostering tensity and balancing emerge. In the case of steel and concrete, the pleasure is largely hedonistic.

THE EFFLUENT SOCIETY

There is money to be made, resulting, most blatantly, in anthropogenic nuclei of particulate matter, carcinogens, metal trace elements, and some six million potentially synergetic chemical compounds disrupting the nitrogen, air, water, and phosphorous cycles. Almost all of the 70,000 toxins now in commerical production are released into our lives without prior study. The Environmental Protection Agency (EPA) estimates that another 50,000 "Love Canals" await discovery across America. What has all this "prosperity" suggested – unlimited potential? Hardly. Instead of increasing human carrying capacity per acre of space, or per ton of substance, this technology has merely become a means of increasing the space required per human occupant and the substance necessary for each consumer.

ECOLOGICAL CRASH

There have been innumerable civilizations and colonies that overshot their limits. Mayan culture disppeared around 900 A.D. from cumulative destruction of its watersheds, resulting in massive topsoil erosion. The destruction of the Tigris and Euphrates agricultural region, the Irish potato famine, competitive wipe-out on Easter Island – such examples have stirring analogies in the animal kingdom. Lemmus trimucronatus on the Alaskan tundra goes through four-year cycles of boom and bust. The lynx-hare, Lotka-Volterra graphs and the classic St. Matthew Island colonizer reindeer crash of 1966, amply lend a degree of devout credibility to the Council of Rome. Nobody likes to hear of limits. When Carter asked for a stepdown in per capita consumption, following his energy bill, Howard Baker, in a television interview, desperately averred, "We'll never give up!"

THE GETTING OF WISDOM

What does a human being need to live? In Lucretius, Virgil, Lao Tzu, Thomas Moore, Thoreau, we are given countless displays of enlightened self-interest, of sufficiency. The size of a door, of a bedroom, has remained basically the same for all people in all times. Why? If there is this bottom line of contentment, is it enough for you and me? I wonder. My own experiences with Bedouins in Sinai, with Tibetan yak herders, with shepherds in the Alps, in Bhutan, with rural folk in Central China, Bangladesh, Alaska, convince me that most inhabitants of the blue planet can survive admirably without vacuum cleaners or recreational vehicles. If we have identified our future in terms of that material matrix to which boredom aspires, it is probably because we are all orphans, living without the context which nourishes the last remaining Hopi on their mesas, and the Gé-speaking tribes throughout the Matto Grosso in Brazil.

LIMITS OF ALTRUISM

What are we willing to forsake? How much can we expect from the present? Over one trillion dollars could be saved were the U.S. to exercise restraint in the following areas: oppulent fashion, redundancy of services, excessive advertising and packaging, consumer fraud, exploitation industries, crime prevention, environmental health costs, unnecessary government regulations, Federal wastage, tax loopholes. The price tag on actual environmental degradation is incalculable.

A moral husbandman, such as Wendell Berry describes, is hardpressed in a market economy. The tragedy of the commons is in fact the tragedy of 20th century rationalism. Yet natural selection favors cooperatives. We have only to examine the !Kung San of Botswana, the semai of Malaya, the Akuri of Surinam, Birhar in Central India, Menpa in Tibet, Sanye in Tanzania, Eskimoes in Northwest Territories; protozoa, sea urchens, herring gulls, planarian worms, goldfish, Hyponomenutac caterpillars, wolves. The rationalists have discounted these exemplars. Lorenz, Ardrey, Tennyson, Thomas Huxley, Herbert Spencer, John D. Rockefeller, have all ascribed to social Darwinism, sanctioning the survival of the fittest, rather than survival of the fit. This is a mistake.

THROUGH THE LOOKING GLASS: ECOLOGICAL TYPES

What are the alternatives in an age seeking substitutes for itself? I would, at the risk of sounding "unrealistic," first draw attention to three groups. The Drukpa of Bhutan partake in the earth with full blushing reverence. Applicable descriptors include wise husbandry, population control through natural birth control and polyandry, sparse cottage industry, modest, diverse agriculture and agroforestry, hunter-gathering. One million people. Their secret? Tibetan Buddhism and isolation. The Qollahuaya of Mt. Kaata in Northern Bolivia. They exhibit the most workable land tenure on earth, according to traditions as old as the potato, which they helped develop. Their formula? A pragmatic, deep understanding of highland ecology; of their surrounding pharmacoepia. The Hushi of the Baltoro, in northern Pakistan. This inaccessible people inhabits the wildest mountain terrain from Detroit to Rangoon. Their amenities are few. Their luxury of lifestyle lavish. What's that? It means human scale.

A SUSTAINABLE IDEA[1]

In the cycle of deflowering, we must begin somewhere. Let us begin with population. Fertility replacement means minimum throughput. The World Bank should get off its high tech loaning kick, and put many more eggs into family planning projects. Less-developed countries (LDCs) would do well to stop hoping for subsidies from the likes of our administrators, and reread Gandhi. Their land is their responsibility. They must manage it for themselves. We must disabuse ourselves of the notion that we can foster a global commons. It would be disastrous. CARE is correct in citing the fact that certain desperate measures are sometimes necessary. Johnson's aid package to India saved millions of lives. Now India must cope for herself. Our coffers are bankrupt. Fifty years ago the Nobel laureate chemist Frederick Sotty saw this and insightfully described a true balanced budget. What you can't afford, you shouldn't deficit finance. Wealth must not be confused with stock, as it so rampantly is today.

Dozens of illustrious blueprints specifying new technological breakthroughs for planet earth have arisen in the past decade. Dennis Hayes, Paul and Anne Ehrlich, Lester Brown, Eric Eckholm, Hazel Henderson, Ralph Nader, Amory Lovins, John Holdren, Garrett Hardin, and the Meadows are some of the well-known protagonists in the mid-century limelight of ecological speculation, authors of the many techno-tractates. They cover politics, economics, engineering, ethics, biology, architecture, sociology. There is no prospect left unattended; no incursion that has not witnessed intensive, if contentious scrutiny. At no other time in its ontology has the mind pondered itself with quite the frenzy, the panic. Indeed, the future amphetamizes all speculation, charging it with an indisputable excitement not unakin, at times, to disaster nostalgia. My first visit to the Pentagon was memorable: Three top-brass were on their hands and knees in some inner sanctum playing with toy, automated tanks. On the walls of the office, disaster movie posters. That particular scenario has always stuck with me.

One billion people, and billions of animals, go to bed hungry every night. Why not try it tonight for yourself. Or better, go a week unsatisfied. Of course you're still better off because your stool is no doubt firm and mellow, free of worms; your clothes are pressed, your shelter warm, your immediate future given to your control. This is the normal incentive of profit control. It does not easily foster true community. Hunger does not, either. In the Sahel, hunger drives the young away from their villages into cities, where they struggle to derive monies from the marketplace and bring them home again: it's the only way they can obtain emergency rations – with cash, which they are not normally accustomed to bargaining with. Coming home to withered villages with a few bucks in their pockets, they eat, and re-populate. Strikingly, population *rises* in the wake of famine. And with population comes all the attendant pressures on the land.

By the twenty-second century, even assuming global fertility replacement, the inherent momentum of population will reach between nine and twelve billion, or nearly three times what it is today. World Food partisans estimate that technology could feed 20 billion people under sufficiently financed governments and by systems of efficient distribution. There are those who would have us believe that 20 billion people is desirable. And if it proves too noisy, why then there will be space colonies to absorb the overflow. No biological analogy can convince the ardently unconverted. Strangely, were nature to adequately portray pandemonium, the staunch mind would never acquiesce, plunging, instead, to its grave ever redeemed, ever inflated with its hubris. As the disaster curve stretches higher, the mind verges towards creationism, that is to say, it was meant to be this way. I'm speaking of asymptotic ethical systems by which values never meet, but glance, momentarily, at the conference table, wince a word or two, then retreat forever into private, indurate darknesses. We haven't yet learned how to talk with each other. Nor do we know quite what to say, given the chance.

LAND

So on one hand you have the techno-tractates. Let's summarize their drifts: In terms of hectares, Wendell Berry has most eloquently called for a return to the small family farmer, citing instances where organic yields from 160-acre plots have nearly equalled, in some cases bettered, massive computerized holdings. Fifty-million hectares of degraded land in India has been identified as suitable for agroforestry. Central Africa's exhausted watersheds could similarly enjoy such rejuvenation. New technologies for integrated pest management, laser contour cropping, genetic engineering, new rotations and manuring; hydroponics; closed (protected) farming; aquaculture; and the paramount germ-plasm banking, all bode of minibreakthroughs. The Red Book identifies at least 25,000 endangered plant species. Desertification, unceasing deforestation, shortening fallow cycles – these desperate patterns are destroying that one-third of the planet allotted to terrestrial beings. The peasant cannot afford fancy technology. *Appropriate technology* is his answer.

RECYCLING

Returnable bottle legislation in the United States would annually save some 500,000 tons of aluminum, 1.5 million tons of steel, 5.2 million tons of glass. The nonenergy companies are seizing upon such savings. But so far, major energy consortia are not entering into the conservation market. As Lester Brown points out, virgin ores still receive preferential rates set by the Interstate Commerce Commission, whereas scrap metals should get the breaks. Millions of serviceable garments made of natural fibers are retired unnecessarily each year. One cannot begin to estimate the costs incurred by fluctuant vogues.

FUEL

Auto fuel consumption could drop precipitously through the implementation of a full range of energy-conscious measures. More efficient Third World cooking stoves could cut firewood consumption by half. Home insulation in this country and Europe could reduce heating and cooling costs by 40%. Munich derives 12% of its electricity from garbage. Paris and Rotterdam are doing the same. Innumerable fruit and creamery wastes can all be converted into fuel-grade alcohol for valuable increases of gas consumption. The transformation of cattle yards into methane-producing energy farms holds abundant promise. The World Bank Third Window loans, UNEP and the Agency for International Development (AID) should research small production units for the Third World. Such appropriate techno-fixes could revolutionize energy usage by providing fuel without sacrificing soil fertility. The food-fuel competition might be averted by the identification of undomesticated plants (i.e., *Euphoriba lathyris*) with high oil contents for planting in otherwise poor soils, and capable of yielding 6.5 barrels of oil per acre. Sunflower seed oil, vegetable oil fuels, African palm oil, water hyacinths, kelp, algal blooms – all possess fuel-conversion possibilities. Sewage enriched warm water can produce several tons of water hyacinths each day which is enough to yield several thousand cubic feet of methane.

WATER

Ironically, the building of big dams not only disturbs the ecosystem, but tends to make nations dependent on one another. It has been suggested that Sadat went to Jerusalem because he realized how easily Israel could destroy the Aswan High Dam, and Egypt with it. The mountainous countries of the world – in Asia and South America – are poor, and in earnest need of the power to be derived from their clean, sustainable water courses. New dams on the Yellow and Yangtze could quadruple China's existing generating capacity. The primary dilemma here is the environmental-cultural trade-off: valuations of wilderness and aborigine culture. Chinese minihydro projects at the community level now account for 33% of China's existing generating capacity. The water is employed in the running of irrigation pumps, grist mills, sawmills, local light industry. In the United States, 50,000 dams for water storage and flood control or recreation are not being used for electrification. Some 9000 good conversion sites have been identified, and such conversion can be construed as local light industry, its profits diffused and energy generation used locally. Water power accounts for more total U.S. energy output than does nuclear. Yet there is massive new opportunity for water power. In Europe, 60% of all rivers have been tapped. While there are more aesthetic, less damaging forms of tapping a river, the essential question persists: Water power versus nuclear power? Water must win out. The first hydro-electricity system is now in place beneath Mt. Everest.

WIND

The exploitation of the sun's kinetic energy on the earth's surface presents superb possibilities. Wind power was first used to lift water in Persia in 600 B.C. Today, the largest wind generator is situated near Ulforg, Western Jutland in Denmark. Boeing prototypes are being developed in Washington State. By 1990, the U.S.S.R. will have built some 150,000 wind turbines.

THERMAL

Seventy kilometers northwest of Lhasa, the Chinese have identified a thermal source and plan to energize Lhasa with it. 65% of all Icelandic homes are heated by underground hot water. California could be drawing 25% of its electricity from geothermal sources by the year 2000. Japan is presently doing so. The U.S.S.R. plans to tap the Avashinski volcano.

SOLAR

Fifty-thousand rooftop solar water heaters existed in Miami alone, during the 1950s, but cheap oil and gas put them out of business. Ten percent of all Israeli homes use solar energy for their heating. In as much as 33% of all U.S. energy consumption is for heating, solar conversions represent an important transition. The photovoltaic array implemented in the Papago Indian Reservation at Schuchuli, Arizona, in 1978 was an unusually successful beginning. Vermont has waived property taxes on residential solar installations. The first all-solar home was erected for commercial purposes in New Mexico in 1981, and all-solar businesses have been springing up in California for several years.

DESIGN WITH NATURE

Ian McHarg, Christopher Tunnard, and Paolo Soleri have each outlined visions of the urban future in communion with environment. The University of Saskatchewan Village House 1 – a passive solar structure – costs but $40 a year to heat. There is no furnace, no scientific chicanery, merely good insulation. In America, the average home costs $680 a year to heat. The world's buildings could use 25% less fuel and electricity in the year 2000 with proper design implementations.

OTHER PROTOTYPES

There are, as Lester Brown indicates, several additional energy-saving methods, including the comeback of draft animals; labor, versus energy-intensive work; Eucalyptus tree plantations in California and thermal gradients in oceanic Florida all offer additional energy grids.

Alternative energy sources combined will possibly yield in the year 2000 some 3635 million metric tons of coal equivalency, or between 25 to 30% of the total needed coal energy. Dennis Hayes sees a 69-quad energy need per year in the year 2000, with 30% (we hope) coming from soft energy paths, accounting for a 20% reduction in throughput.

Is all of this anachronism or guide? Remember Perrez Alfonso, with his 1936 Singer Sports coupe parked on a pedestal in his garden? He founded OPEC, and believed it to be the leading ecology group in the world. He knew that it force upon the West the need of conservation. How have we come to this massive watershed of alternate futures?

THE ECOLOGICAL CONTRACT

The social contract arose to divide the spoils of the great frontier and of the industrial revolution. Originally, it overthrew medieval political kingships. But such an agreement was based upon the cornucopian myth of continuing abundance. Nature, in such a scheme, was held external to politics, taken for granted. The Romantic impulse after Arcadia stemmed from the sudden, tumultuous upwelling of recognition: Nature, and

Rousseau's *l'homme sauvage*, had fallen to the ax of manifest destinations. The artist's eye seized earnestly upon glowing embers of some prehistory which seemed, still, to survive in isolated remnants of Savoy, the Rocky Mountains, the Hindu Kush, in Melanesia. The passion to seize upon these visions of paradise are seen in the paintings of Claude, Salvatore Rosa, Hercules Seghers, Watteau, and later, in the landscapes of the Royal Academy and British Institution, and of Friedrich in northern Germany. The Hudson River served to focus the American spinoff of these efforts, and after the river, it was the West.

SCHIZOPHRENIC DESIRES

The two modes – humanism and economics – exerted a prodigious stranglehold, from the time of Pizarro's conquest of the Inca to the massacre of Native Americans, as Europeans and "gentleman farmers" idealized the very *indigenes* who were systematically vanquished. The New World saved Europe. Whereas Aristotle had advocated slavery as civilization's only answer to scarcity, Jefferson went out and bought the Louisiana Territory, tripling the size of America. We needed a spread, said he, wide enough to ensure liberty. Indeed, democracy demands carrying capacity. Jefferson envisioned this magnitude in terms of rural, agricultural, populist hard work. Hamilton saw the urban, elitist, commercial banker as farmer. Together, in a more or less amicable squabble, the wealth of America was chipped away. Any social problems were handled by substituting economic growth for political principle. Ironically, even some of the artists got rich in this tide of double-bind interest. The painters Albert Bierstadt and Frederick Church inspired, then capitalized upon, the emergent euphoria in Washington and in galleries throughout the East.

ERA OF ABUNDANCE

Tertullian (160 - 230) first castigated technology as unnatural. Only a few monks heard him. Saint Simon (1760–1825), high priest of the techno-fix, did not hear him. His own theories corresponded with the writings of John Locke and Adam Smith: With consumers and producers acting "rationally" to maximize their own gain, the market would automatically allocate resources with greatest efficiency and generate the maximum individual and social prosperity. This expectation for growth was written into U.S. domestic, and eventually foreign, policy. Marx would celebrate the coming enormous productive force so that the bourgeois would overthrow feudalism, and resulting scarcity and inequity would foster the triumphant ascendancy of the proletariat, or ruling class. Edmund Burke (*Reflections on the Revolution in France*) was the last great spokesman for the premodern point of view in political economy. He was skeptical of progress, recognized the need to accept limits, and saw society increasingly ruled by the amoral capitalists, "sophisters, economists, and calculators" (what Jefferson thought of lawyers); J. S. Mill was as distressed with the "trampling, crushing, elbowing, and treading on each other's heels." He called for the stationary state economy and was reaffirmed in that outcry by Dickens' Coke Town. Thoreau's own choppy, sentient, effusion admits to the confusion of the era. While he relished the solitude and naturalism of Walden Pond, he longed for the daily whistle of the train on the water's far edge.

RESTRAINT?

The conversion ethic arose in Congressional debates over the first preserves. Painters and poets were largely responsible for this burgeoning tolerance. Yellowstone, however, did not come into existence because its beauties were deemed inviolable. Yellowstone was simply considered wasteland, economically unexploitable. The rise of western photography served the aesthetic theories of John Ruskin as much as it did the lobbying of John Muir. Roosevelt's 1908 National Conservation Commission, and the sudden intrusion of auto traffic in Yosemite, the land-grab at Niagara, bespeak of the democratization of wilderness, leading thematically into its own madness. While wilderness conservation made gains in the writings of Henry Beston and Aldo Leopold, larger political forums were at work to address the concept of limited and unlimited resources. Hoover's Research Committee on Social Trends, Roosevelt's National Planning Board, the Paley Commission under Truman, and Resources for the Future Committee all led unequivocally to Rachel Carson and Aurelio Peccei. It was not until 1969 that the environment enjoyed government support in any superior way, however. Executive Order 12114, and NEPA, called for a global commons and for global environmental impact statements. It was only under Carter's administration that any group effort in government was undertaken to assess the world environmental situation. This massive enterprise, resulting in The *Global 2000 Report* comprised Federal participants who had never met one another. Government concern for nature is based on a grand deferral system. Profits now, concept later. No one is guiltless, save those few independently impoverished, landed families – in Scotland, northern California, Israel, Sikkim – who have provided their own nourishment and shelters in the absence of utility networks, aggregate spending, and resource depletion. And indeed, such cooperatives represent true health. But the governments of most countries do not want to know about it. This is perhaps too harsh: Governments are not 'bad guys.' Faced with tightropes, under pressure from ever broadening constituents and interest factions, they tread from nightmare to nightmare.

Until conservation and development are shown to be compatible; until nonmonetary indicators of conservation performance can be chosen for inclusion in any national accounting system; and until each individual addresses his own basic wants on the plank of altruism, there will be no substantial change. The innumerable global conservation conventions since Stockholm in 1972, leading to the World Conservation Strategy, have made a powerful statement but little more. Seen in the context of a century, the first intimations in the wake of park nationalization – concessionaires' disease – have not been checked. We now recognize that the government can open up prime lands for multiple use, as their mandate approves. And while worldwide acknowledgement of pollution, and chemical toxins has been achieved, Montana's fish and game authority can still ignore endrin counts in waterfowl, and DDT continues to be applied in most parts of the world. Pollution standards are ever eroded, increasing laws are legislated, thus assuring diminished freedom without guaranteeing reciprocal protection for the environment. There are no labor unions fighting for nature, despite decades of National Geographic's religious pictorials.

WAR MENTALITY[2]

Amid this rampant decline in quality of life for our species, the war mongering escalates – a clear reaction to stress. The League of Nations failed, and while the 1922 Washington Armament Conference declared a moratorium on naval shipbuilding and the 1928 Briand-Kellogg Pact renounced war, the German battleground taught us that war was like a perennial. Still, America had tried, as C. Maxwell Stanley has clearly elucidated. Among other things, the U.S. instituted the Marshall and Baruch Plans, Part IV Technical Assistance, and, previously, working out the broad outlines for the United Nations at Dumbarton Oaks. Today, human, electronic, and satellite intelligence reports are summarized by DIA and CIA logicians, before being presented under the title of National Intelligence Estimate to the Joint Chiefs, whose chairman issues the National Posture Statement, passing on recommendations to the Office of Under Secretary of Defense for Research and Engineering, to the Department of Energy for Nuclear Arms, or to the Defense Advanced Research Agency, which in turn hires its subcontractors. The 1981 posture report cites figures indicating the Soviet overkill superior to our own. We create such overkill to absorb first strike. How did Hitler and Stalin so catalyze this staggering psychology, and why is there no international Sanhedrin to defuel it?

U.N. REFORMS AND DISARMAMENT

United Nations bipartisan consensus is required on the following principle: the terrorizing of other nations with a show of military wherewithal is no longer a tolerable form of diplomacy. Under conditions of reciprocity, there must be a moratorium on all further testing and the General Assembly must adopt a comprehensive test ban.[3] In so doing, the United Nations would finally prove itself capable of acting, rather than merely reacting. Leo Szilard's book *Voice of the Dolphins* raised the issue of a third-party form of resolution. It's time for countries to listen, to cede their arrogance over to a higher authority. Greater power must be given the Committee on Disarmament. Nuclear free zones must be examined (e.g. the Treaty of Tlatelolco). Nine nations export 95% of all arms transfers. If they could agree to limit such transfers, as was accomplished by France, the United States and United Kingdom in the Middle East from 1950 through 1956, the first juggernaut would have been cracked. The present status of the Arms Control and Disarmament Agency is disgraceful: a meager annual budget of 16.5 *million*; compare that figure to the 56 *billion* for military research and development. "There is no point for exporting countries to expect nonweapons countries that see themselves under grave threat to resist the temptation to go nuclear," says Eugene Rostow, the agency's former head. The Saudis got AWACs to protect their oil fields; Brazil and Argentina defer conflict because both share a critical dam on the Paraná. Environmental resource requirements supercede ideological abstractions, and it was always so. War is the outgrowth of ecological conflict, the *biosphere* versus the *hypothalamus*. But the point is not mere reductionist finery. Getting our environmental priorities straight implies a solution to all other problems. Why, then, do we avoid the issue?

MUDDLING THROUGH

A muddling-through policy is policy by default, an administrative device in this country – in all bureaucracies – for divorcing common sense from need. A goal-determined method of decision making may not be fully realizable in a world increasingly constrained, where ethics are sacrificed for efficiency, which in turn glosses over information, rendering plain speaking and plain thinking a thing of the past. We have taken the position that there can be no common interest beyond what muddling through induces: an "adhocracy." We continue to react, rather than act, with increasingly dangerous *tensity*. Neither "side" is willing to give an inch. A show of sanity is construed as a sign of vulnerability. The scenario for this is political overload – problems growing faster than the wherewithal at large to handle them. Fragmented administrators with specialized tunnel vision foster lagtimes, inter-agency conflict (e.g. land conversion, riparian rights, EPA jurisdiction); while Federal, state, and local governments attempt to independently handle elements of the law – but such political delinations have little or no bearing on ecological regions. Furthermore, courts are unaccustomed to environmental litigation and to the kinds of new statistical and epidemiological evidence being wielded as proof. Courts can not defend the home. How can they begin to defend the planet? People won't listen to reason, or to ethics. What concerns them is money.

ECONOMICS

The earliest physiocratic economic theorists explained buying and selling according to natural law in which Mother Earth was the source and the only lasting recourse for all value. Money was perceived as sterile, and the reproduction of interest was rejected on the greater basis of plant and animal reproduction. John Keynes changed all that through the New Deal, and a world engine deus ex machina that brought Americans out of the Depression. The International Monetary Fund and the World Bank came about in the wake of the Marshall Plan and the Bretton Woods order. With the lending, the dollar convertibility, and the emergence of surplus, we were back to the seventeenth-century notion of infinite growth. But the order disintegrated in 1971, special drawing rights were introduced, floating rates, and the number of national currencies mushroomed to nearly 130. There was no longer any standard because the throughput of capital had exceeded the reasonable rate of exchange on the part of natural resources. In other words, human consumption, aggregate wants, could only be accounted for by the production of new money, in lieu of new nature. True, the gains in fossil fuel, agricultural, and fisheries production were dramatic; investment capital was aggressive, profits were high. But there was no confidence; there was no stability; and then there was even less confidence. In December 1973, the Arab oil embargo shattered all hopes for a continuance of such abundance. Later oil shocks drove home the message that had been ignored for a century. OPEC may have incurred the wrath of western consumers but succeeded in introducing a mighty conservationist lesson – self-sufficiency. The commodity cartel offers the Third World a collective opinion to manage its resources, upgrade the quality and quantity of its local work force, and thus negotiate trade agreements from a more fully insulated position. Trade has long been viewed as the key means of cementing differences between nations. Where trade is equitable, balance, production will be so as well. We see this in the declining end-user's electrical grid in western nations, which follows, and triggers, a reduction in Middle Eastern oil procurement.

THERMODYNAMICS

Photosynthesis does not maximize throughput, and is only 1% efficient in terms of energy production. In fact, 99% of all solar radiation goes to maintaining the biosphere. The first and second laws of thermodynamics reaffirm that energy can not be reused or destroyed. All energy disperses as increasingly inaccessible heat waste. Matter cannot be created, only recycled at tremendous energy loss or high entropy. N. G. Roegen observed that entropy becomes a fine way of seeing the real basis of economic scarcity. It will never pay mankind to run a completely technological world, because the life support system of nature is too inexpensive. Yet try to recreate a beaver lodge and pond! The entropic flow of nature, characterizing thermodynamics (first perceived in the heat-engine efficiency studies of Sadi Carnot, and the balanced equations of Antoine Lavoisier) explains the great migrations of races, of pastoralists, and clarifies the process of throughput by which a market operates. Borrowing from physics, Herman Daly's three visions of the economic future have revolutionized ecological criteria for sustainability. In Daly's steady-state process, a stock should not grow beyond where its marginal cost is just equal to its marginal benefits. At some point, economic growth becomes antieconomic. The inflationary spiral is not given to temperate responses, but rather economic panic, overshoot, to use Catton's term. New measures of wealth must come into our understanding, for the gross national product (GNP) is inadequate, and if anything, a gross national cost. NEPA, Section 102, demands an impact statement for every government-related project likely to impact. But Section 102 is shot through with loopholes. Both Tellico Dam and the Alaskan pipeline were passed through devious circumvention of the NEPA strictures, the bias firmly with development. A fiscal budget, to be balanced, must be balancing in terms of natural resources. The natural resource units (NRU) theory has been propounded: The government allocates such resource units in recognition of a finite number of units. The consumer is free to juggle his units, to sell them, spend them all at once; but the units are allocated to prevent the over-all consumption of units from exceeding that which is environmentally sustainable. The limits to growth paradigm was not programmed to reveal the sustainable point where growth must stop. Cost-benefit marginality does just that in the steady-state plan.

NURTURANCE

Where does this leave us? With marasmus, for one. So many infants died in hospitals in the early 1900s because they had no mother's love. What does this have to do with economics, the United Nations, war, ecology? The mother–child landscape is the first landscape in our life. Our developing psyche and neuropsychology, during the critical periods of birth, during the first 3 days, the first 7 years, dictate the future of our species. Freud told us that human anxiety begins during the infant's sense of helplessness. Man will do anything not to be doomed to himself. A child needs, from the beginning, a sense of community, of nurturance. The child's literal, valueless contact with nature is easily transformed into a strong, ego centeredness. As an adult, when forced to weigh literal values, he'll always choose progress. Introducing the child to nature must be seen as crucial to humanity in the coming era of radical will. Otherwise the child, having missed out on any truly coherent contact with the natural world – sustained contact – is likely to seek order in his universe by turning only to machines, as Paul Shepard has pointed out. Much has been written on deschooling, the return in education to human scale: the work of Roger Barker versus the Conant Report – small, intimate clusters versus the assembly line. The 1966 U.S. Office of Education survey indicated that the major classroom determinants were "the attitudes of student interest in school, self-concept, and sense of environmental control." What has not been written is of education on a farm, a Hopi Mesa, in an Ituri rain forest. Mystery, ritual, initiation: the bringing back of reverence and mysticism to temper street-smartness: Peanuts, Plotinus, and the Planet.

The parent–child relationship is the pivotal point in the ecological revolution. The year 2000 will see today's youth in a position to alter the future. It has been said before. But I doubt whether so many people ever cared to think about it so much as now. If it's true, then we'll *do* something. If it's untrue, I, for one, am uncertain whether anyone else is listening. We have fashioned a world in such a manner as to lend our children ardent preoccupations. They are not necessarily *good* preoccupations. What is worth doing, after all, when you're a kid? Playing computer games, or making out beside the old swimming hole, following turtles, gazing after wild horses, rolling down hillsides?

Liberals in this beleaguered hour speak out in defense of realism first, compassion second. They have little choice if they are to regain the public's vested interest. Civilizations rise and fall, like anonymous stars. But mothers, by being compassionate with their children, encouraging natural values, exposing their flesh and blood to the earth cycles, *are* realistic. And until this message is driven home, there can be no true ecology, no revival of the human spirit. I'm not even sure what such a revival would mean. It would taste of grain, smell fresh, like muggy tropical breeze, or the cool, irradiated night of alpines; it would sound strange and sibilant, like romantic love or poetry; and it would feel secure, almost mucousy, for the number of friends around the place. Overhead the stars would shine without the light-diffusing effect of cities. The gut would feel fine, the heart resonant. In the morning, there would be earnest work to be done. Ideas would be dirt stained, after Eden.

NOTES

[1] For much of the energy and statistical data sections proceeding "A Sustainable Idea," I am indebted to Lester Brown and his dozen, exceptional World Watch Institute publications. In addition, I have relied upon Brown's *Building A Sustainable Society* (W.W. Norton & Co., 1981), The Council on Environmental Quality, *Environmental Quality 1982* (Washington D.C.: U.S. Government Printing Office, 1983); and the *World Conservation Strategy 1980*) (IUCN, Gland, Switzerland).

[2] For the sections on war, performs and economics, I have used C. Maxwell Stanley's brilliant work, *A Guide to Survival – Managing Global Problems*, The Stanley Foundation (Muscatine, Iowa, 1979), pp. 37–128.

[3] C. Maxwell Stanley, p. 53.

Ecological Consciousness and Paradigm Change

George Sessions

George Sessions teaches philosophy at Sierra College in California. A mountaineer, Sessions' main concern since Earthday I has been the development of a new ecological philosophy. He has written a number of papers in the area of ecophilosophy and has edited an ecophilosophy newsletter.

The prevailing mood among avant garde intellectuals and many others throughout the world is for massive social paradigm change – for radically new directions for humanity. The specters raised by Huxley's *Brave New World* and Orwell's *1984* have increasingly become social reality. These prophets of dystopia and social disaster warned primarily of the "diminishment of man" brought about by increasing technological-political manipulation together with the threat of continued world-wide war. The new ingredients in the witch's brew are the interlocking problems of human overpopulation, mass starvation, and overall ecological disaster, although Huxley was one of the first to warn of these as well.[1] As we are now beginning to realize, the key to contemporary ecological consciousness is to see the diminishment of man and the diminishment of the planet and its nonhuman inhabitants as essentially one and the same problem.

Many people in industrialized countries are aware of the host of contemporary environmental problems although they often fail to see them as interrelated; air and water pollution and the death of the oceans, toxic waste disposal, the chemicals in our foods, the rise of environmentally related cancers and other health problems, nuclear power and the "energy crisis," the loss of prime agricultural land to urban sprawl, the destructive land practices of modern industrial agriculture, the rate of destruction of forest ecosystems in Central and South America, Alaska, and over most of the planet, the likelihood of dangerous global climatic change as the result of increased pollution and energy use, dwindling resources, the threat of world-wide famine. People are less aware of the rate of species extinction and overall habitat destruction; *The U.S. Presidential Global 2000 Report* estimates that 2 million species of plants and animals will become extinct in the next 20 years, and 5 billion new acres of desert will have been created. All of this is the mostly familiar litany of environmental ills for which specific "solutions" – usually technological – are sought.

Fewer people, usually those with a basic grasp of the scientific principles of ecology, realize that we are unraveling the integrity of the world's ecosystems – we have been literally destroying the very biosphere itself at an exponentially increasing rate since World War II.[2] But even those with a sophisticated knowledge of ecological principles seem unable to face the implications of Barry Commoner's Third Law of Ecology that "Nature Knows Best." Commoner explains this as meaning "any major man-made change in a natural system is likely to be *detrimental* to that system." Commoner further points out that this law is especially unpalatable to modern technological man because it challenges the idea of the "unique competence of human beings."[3]

As many historians who have researched the ecological problem now point out, the major thrust of Western culture, of both our Greek and Judeo-Christian heritage, has been to assert the uniqueness of humans, to emphasize our separation from the natural world and other life forms, and to see our role as having dominion over the rest of nature. The Australian philosopher-historian John Passmore has traced the historical development of Western views from "man as despot" and reckless exploiter, to more recent view of humans as "stewards" over the earth's resources (resource conservation), and of humans as "humanizing" and "improving" the earth (resource development).[4] The biologist Rene Dubos seems especially enamored with the idea of man's role as improving on natural systems.[5] Since *all* of these dominant Western views of man's place in nature fly in the face of basic ecological principles and laws, we can begin to see why ecology has been called the "subversive science."[6] Ecology shows us that the basic assumptions upon which the modern urban-industrial edifice of Western culture rests are erroneous and highly dangerous. An ecologically harmonious social paradigm shift is going to require a *total* reorientation of the thrust of Western culture.

I

The whole issue of Western man's arrogance toward the rest of nature was recently highlighted by the controversy that arose over the captive breeding program to "save" the California condor from extinction. With only about 30 of the birds left, the U.S. Fish and Wildlife Service, with the blessings of the National Audubon Society, instituted a captive breeding program that was to begin by trapping the birds and outfitting them with tiny radio transmitters. Later, nine of the condors were to become breeding stock at the San Diego Zoo. This program was supported by most of the major conservation organizations and by biological research teams. The leading condor expert, Berkeley zoologist Carl Koford, and Dave Brower of Friends of the Earth disagreed. Koford thought the birds could make a comeback without direct interference in the birds' affairs by scientists if the habitat were enlarged and restored. Koford pointed out: "With a cage-bred animal, you may be able to preserve genes and cells and organs. But a wild animal is much more than these. Can you preserve its wildness? Can you preserve the complex cultural heritage that animals in their natural habitat pass along from one generation to the next?"

But the program went ahead as an example of "good" scientific/ technological "hands-on" wildlife management until disaster struck. World-wide attention was focused on the program in June 1980 when one of two condor chicks hatched that year died from trauma while being weighed and measured by biologists. It did not take long for the deeper philosophical issues to be raised. Mark Palmer of the Sierra Club asked, "However well-intentioned it may be, isn't captive breeding yet another manifestation of our presumed right to manipulate all life on this planet for our own interests?" In his article on the condor, Robert Reid suggested that

> *Preventing the extinction of a species might further human interests by seeming to demonstrate the triumph of compassion, and possibly technology, over the grim forces of evolution. And it could furnish self-satisfying, though no doubts self-deluding, evidence that the human excesses which may have driven the species to the brink of extinction in the first place were not really so excessive after all.*[7]

He further added that if the condors die we will have lost two things we covet very much: "confidence in our ability to manage the affairs of the world, and condors."[8]

And so it goes. Most of the large mammals of the earth and sea are on the endangered species list as a result of habitat loss and/or over-hunting. Zoos now have the excuse to continue to capture and confine them as part of a "scientific" breeding program and to maintain the genetic diversity in "sperm banks" – a modern "Noah's Ark," although without hatural habitat they will forever remain captive curiosities with no possibility for natural selection.

The future possibilities for non-captive wild plants and animals and for the natural evolutionary processes on this planet are indeed grim. Nairobi-based wildlife expert Norman Myers has outlined the situation in his book *The Sinking Ark* (1979). More recently, he has pointed out that

> *If we consider all species on Earth, and the rate at which natural environments are being disrupted if not destroyed, it is not unrealistic to suppose that we are losing at least one species per day. By the end of the 1980s, we could be losing one species per hour. It is entirely in the cards that by the end of this century, we could lose as many as one million species, and a good many more within the following few decades – until such time as growth in human numbers stabilizes, and until growth in over-consumerist lifestyles changes course.*[9]

Myers has proposed a system of "triage" for wild species. It is impossible, he argues, to even begin to save them all. Certain species should be selected when there is some hope for success.[10]

The Canadian naturalist John Livingston has tried to come to grips with the tremendous failure to protect wildlife during this century. The arguments used have all been "rational," meaning based upon narrowly conceived human interests. Plants and animals have been treated, in theory and practice, as only actual or potential "resources" to be "consumed" in various ways. Without a major change of consciousness, a profound and intimate sense of interrelatedness with nonhuman nature, Livingston sees no hope for beginning to turn the situation around.[11]

Until recently, modern people who considered themselves "enlightened" on the human–nature relationship have thought of themselves as "conservationists." John Livingston defined *conservation* as "the care of 'natural resources' and their protection from depletion, waste, and damage, so that they will be readily at hand through perpetuity." John Passmore defined it as the "saving of natural resources for later consumption."[12] Out of the Progressive movement at the turn of the twentieth century, as a result of the influence of Gifford Pinchot on Teddy Roosevelt, the modern version of ulilitarian Resource Conservation and Development was born. The "wise use" of "resources" would be achieved through the "rational efficient scientific–technological management" of nature for the benefit of "the greatest number of humans." Reckless exploitation would cease and social justice would be achieved. Pinchot became the first director of the U.S. Forest Service and, in keeping with his anthropocentric orientation, announced that there are "just people and resources." The rise of the "resource expert" (the scientific manager of nature) soon followed. Resource management programs were established and expanded rapidly in colleges and universities throughout the world to supply the new demand for scientific foresters, agricultural experts, wildlife experts, range managers, soil scientists, and so forth, to industry and to government resource and development agencies.[13] From this point onward most of wild nature, from forests to wildlife, were to be treated much the same as a domestic field of corn, to be managed and "harvested" by humans.

It is surely no coincidence that humans would also come to be looked upon as a "resource" to be managed in the best interests of the emerging urban–industrial society. The shift from people to personnel (and consumers) to which modern "scientific management principles" are to be applied for more efficient production of commodities is but the flip side of the mentality and consciousness which see nature as but a resource to be managed and manipulated for the benefit of those in power. As Bill Devall points out, we now have programs in our universities to train "leisure management experts" and "wilderness management experts"; "experts" who are largely in the business of managing people.[14]

Loren Eiseley, in describing Robinson Jeffers' poetry, said: "Men feel, in growing numbers, the drawing of a net of dependency against which something wild in their natures still struggles as desperately as trapped fish in a seine."[15] Perhaps as we look into the eyes of captive wild animals in zoos and watch their nervous pacing or boredom, our attitude changes from one of superiority to empathy as what Desmond Morris calls "the human zoo" closes in around us. It is no wonder that Theodore Roszak sees the recent dramatic rise in the movements for self-discovery and personal autonomy as paralleling the equally dramatic rise of the contemporary environmental movement. Roszak claims that "the needs of the planet are the needs of the person. And, therefore, the rights of the person are the rights of the planet."[16]

II

The rejection of the sterile lifestyles and the mindless goal of consumerism and endless materialistic progress of the emerging urban–industrial society by the beatniks and hippies of the 1950s and 1960s, together with their experimentation with the spirituality and self-knowledge techniques of Zen Buddhism and other "esoteric" religions, anticipated the contemporary movements for self-autonomy of which Roszak speaks.[17] Related events led to the emergence in the late 1960s of the contemporary environmental movement and to a shift in philosophical perspective "from *conservation* to *ecology*." The science of ecology provided a unified and interrelated view of nature (including humans) that had been lacking from the discrete problems approach of Resource Conservation, while at the same time exposing the inadequacies of the resource approach. An ecological awareness of interrelatedness, as we have seen, is *subversive* to an exploitive attitude and culture and, if pursued deeply enough at the experiential level, results in what the literary critic Leo Marx called, in 1970, an ecological consciousness or "ecological perspective":

> *that man is wholly and ineluctably embedded in the tissue of natural processes. The interconnections are delicate, infinitely complex, never to be severed. If this organic (or holistic) view has not been popular, it is partly because it calls into question many presuppositions of our culture. Even today an excessive interest in this idea carries, as it did in Emerson's and Jefferson's time, a strong hint of irregularity and possible subversion. (Nowadays it is associated with the anti-bourgeois defense of the environment expounded by the long-haired "cop-outs" of the youth movement). Partly in order to counteract these dangerously idealistic notions, American conservationists often have made a point of seeming hard-headed, which is to say, "realistic" or practical. When their aims have been incorporated in national political programs, notably during the administrations of the two Roosevelts, the emphasis has been upon the efficient use of resources under the supervision of well-trained technicians. . . . In this sense, conservationist thought is pragmatic and meliorist in tenor, whereas ecology is, in the purest meaning of the word, radical.*[18]

Marx found the ecological perspective in much of the American pastoral writing (Cooper, Emerson, Thoreau, Melville, Whitman, and Twain), which had its roots in the Romantic movement. This radical awareness of ecological interpenetration and a "sense of place," he claimed, often produced "a kind of visionary experience, couched in a language of such intense, extreme, even mystical feeling that it is difficult for many readers (though not, significantly, for adherents of today's youth culture) to take it seriously."[19]

By the early 1960s, environmental degradation had become so severe in America as the result of post-World War II economic growth and development that Interior Secretary Stuart Udall, under Kennedy's administration, was about to launch the third major conservation movement in this century. A number of events helped turn this into a questioning of philosophical assumptions. Perhaps the most important event was the impact caused by Rachel Carson's book, *Silent Spring* (1962). Ms. Carson, a trained biologist with a highly developed ecological consciousness and a gift for writing, called attention to the ecological damage resulting from the careless indiscriminate use of pesticides, principally DDT. This had the effect of raising serious doubts concerning the wisdom and competence of technological experts in their efforts to manage the resources responsibly and safely. She may have infuriated the pesticide industry even more with her philosophical challenge to the very idea of managing Nature: "The 'control of nature' is a phrase conceived in arrogance, born of the Neanderthal age of biology and philosophy, when it was supposed that nature exists for the convenience of man."[20]

Secretary Udall's book, *The Quiet Crisis* (1963) was somewhat upstaged by the tremendous shockwaves resulting from *Silent Spring* but it too was a national best-seller. In it, he outlined the American "conservation" crisis and thinking (what now, thanks to ecology and Ray Dasmann, is called the "environmental" crisis), from Thomas Jefferson, through George Perkins Marsh and Thoreau, to Pinchot and John Muir. This was clearly a transitional book. Although Aldo Leopold's ecological ideas were barely discussed, ecological thinking had begun to penetrate the Interior Department. Udall claimed "If asked to select a single volume which contains a noble elegy for the American earth and a plea for a new land ethic, most of us at Interior would vote for Aldo Leopold's *A Sand County Almanac*."[21]

A significant issue which Udall raised was the difficulty he experienced in writing this history of American conservation. The lack of historical and political scholarship in this area was scandalous (pp. 203–226). The crucial issue of the human/nature relationship had largely been ignored by the academic establishment. It was a "hidden variable" in American life left to the province of a few literary "nature writers" and mystics, and to relatively small "special-interest pressure groups" such as the Sierra Club. Tucked away in this ignored issue was, of course, the ticking time bomb of the global environmental crisis together with the most basic assumptions of Western culture (what sociologist Riley Dunlap has called the "human exemptionalist paradigm"). The political scientist Victor Ferkiss now sees the whole issue of *technology* as another explosive "hidden variable" in American life.[22] No wonder most of the intellectual establishment and the general public were so taken by surprise by the dimensions and severity of the environmental crisis and why, after all these years, the majority still seen unable or unwilling to face the real issues squarely. Even most conservation organizations and leaders were not prepared for the full implications of the Age of Ecology. One outstanding exception was David Brower, executive director the Sierra

Club from 1953 to 1969, who has been described as the leading "conservationist" in America. Brower was forced to resign his post as his deepening ecological consciousness and increasingly military stands came in conflict with more conservative Club leaders, whereupon he immediately formed the more radical, ecological Friends of the Earth. A perusal of the *Sierra Club Bulletin* and other Club publications from 1953 to 1970 provides an illuminating chronicle of the "greening" of Dave Brower.

Another very important and sensitive philosophical issue Udall raised concerned changing attitudes of modern Americans toward the nature religions and "land wisdom" of the American Indians. It was ironic, Udall remarked, that

> *today the conservation movement finds itself turning back to ancient Indian land ideas, to the Indian understanding that we are not outside of nature, but of it . . . In recent decades we have slowly come back to some of the truths that the Indian knew from the beginning; that unborn generations have a claim on the land equal to our own; that men need to learn from nature, to keep an ear to the earth, and to replenish their spirits in frequent contacts with animals and wild land. And most important of all, we are recovering a sense of reverence for the land . . . Within a generation (of white settlement of America) the wildness would begin to convert some of their sons, and reverence for the natural world and its forces would eventually sound in much of our literature, finding its prophets in Thoreau and Muir.*[23]

What was prophetic and fascinating about this new interest in, and appreciation of, the "land wisdom" of ancient "primitive" American peoples appeared again to be a growing uneasiness about the modern scientific/technological Resource Conservation and Development approach to nature, and the Western cultural ideas of man's dominance of nature and perpetual progress. The Hopis prophecized in the 1600s that the whites would come and destroy the land and themselves, according to anthropologist Stan Steiner. The Indians would try to tell the whites how to live in peace and harmony with the land, but the whites would not, or could not, hear. Steiner explains the basic Indian philosophy of the sacred Circle of Life: "In the Circle of Life every being is no more, or less, than any other. We are all Sisters and Brothers. Life is shared with the bird, bear, insects, plants, mountains, clouds, stars, sun. To be in harmony with the natural world, one must live within the cycles of life."[24] Steiner sensitively discusses the movement on the part of many American Indians to return to the "old ways" of life and to defend their land against ever-increasing industrial destruction.[25]

In addition to the writings of Carson and Udall, the paper read by historian Lynn White, Jr. to the American Association for the Advancement of Science (AAAS) in December 1966, also caused an immense stir. White argued that orthodox anthropocentric Christianity must assume a large share of the responsibility for environmental crisis by desacrilizing nature, actively encouraging its exploitation, and by promoting a nonecological world view which sees humans as separate from, and superior to, the rest of nature. White also claimed that modern secular humanist ideologies such as Marxism, positivism, and American pragmatism, which have largely shaped the values and goals of modern urban-industrial societies, failed to emancipate themselves from Christian anthropocentric ideas such as man's domination over nature, and a belief in perpetual progress for humans. White concluded that

> *more science and more technology are not going to get us out of the present crisis until we find a new religion or rethink our old one (for they are permeated with orthodox Christian arrogance toward nature) . . . We shall continue to have a worsening ecologic crisis until we reject the Christian axiom that nature has no reason for existence save to serve man.*[26]

White's religious solution was to return to the Christian views of Saint Francis: "the greatest spiritual revolutionary in Western history." Francis tried to derail the Christian drive for dominance and power over nature by preaching poverty ("voluntary simplicity") and proposing a different view of man's relationship to the rest of nature which is remarkably similar to the Indian Circle of Life:

> *Francis tried to substitute the idea of the equality of all creatures, including man, for the idea of man's limitless rule of creation . . . The key to an understanding of Francis is his belief in the virtue of humility — not merely for the individual but for man as a species. Francis tried to depose man from his monarchy over creation and set up a democracy of all God's creatures . . . I propose Francis as a patron saint for ecologists.*[27]

Coupled with these explosive challenges to the foundations of Western cultural assumptions, the backwater biological science of ecology, developed mainly to study nonhuman plant and animal communities, suddenly was elevated to a new status. It was rediscovered that man was an animal too, thus "part of the ecology" and subject to the same principles and laws as nonhumans. This awareness was chilling given what we were doing to the earth. The ecologists LaMonte Cole, Barry Commoner, Paul Ehrlich, and Garrett Hardin, along with Rachel Carson, became cultural heroes overnight (at least to the young), and the stage was set for a very tumultuous Earth Day I, 1970.

III

The history of predictions of environmental disaster and challenges to the Western man–nature view of dominance go back much further than the 1960s. In 1803, Parson Thomas Malthus argued that human population growth would exponentially outstrip food production resulting in "general misery," but his warning was ignored by the rising tide of industrial–technological optimism. The American geographer George Perkins Marsh (*Man In Nature*, 1864), was the first to warn that modern man's impact on environment could result in extinction of the species.[28] The population–resource–environment crisis sketched by Paul Ehrlich in his highly influential *The Population Bomb* (1968) had largely been anticipated by the ecologist William Vogt (*Road to Survival*, 1948). Also in 1948, the British philosopher Bertrand Russell pointed to the dangers of valuing science primarily as technological power over nature, and warned of the possibility of "vast social disaster" resutling from the anthropocentric power philosophies of Karl Marx and John Dewey's American pragmatism which "tend to regard everything nonhuman as mere raw material."[29]

We know that Saint Francis (1181–1226) tried to divert Christianity from anthropocentric dominance to the old animistic–pantheistic position of biocentric equality ("a democracy of all God's creatures"). In the mid-nineteenth century, the German philosopher, Arthur Schopenhauer pointed to,

> *the unnatural distinction Christianity makes between man and the animal world to which he really belongs. It sets up man as all-important, and looks upon animals as merely things . . . Christianity contains in fact a great and essential imperfection in limiting its precepts to man, and in refusing rights to the entire animal world.*[30]

Victor Hugo (1802–1885) called for a new ethic: "In the relations of man with the animals, with the flowers, with the objects of creation, there is a great ethic, scarcely perceived as yet, which will at length break forth into the light and which will be the corollary and complement to human ethics." Albert Schweitzer pointed out that Descartes, Kant, and Western ethical systems in general were anthropocentric. Drawing upon Jainism and other Eastern religions, he arrived as his Christianized version of "Reverence for Life" in 1915. Schweitzer's views have been very influential for the contemporary animal rights and animal liberation movement.[31]

The Harvard philosopher George Santayana scathingly attacked the anthropocentrism of Western philosophy and religion (the "genteel tradition") in a speech at the University of California, Berkeley, in 1911. He commented that

> *A Californian whom I had recently the pleasure of meeting observed that, if the philosophers had lived among your mountains their systems would have been different from what they are. Certainly very different from what those systems are which the European genteel tradition has handed down since Socrates; for these systems are egotistical; directly or indirectly they are anthropocentric, and inspired by the conceited notion that man, or human reason, or the human distinction between good and evil, is the center and pivot of the universe. That is what the mountains and the woods should make you at last ashamed to assert.*[32]

While Calvinism saw both man and nature as sinful and in need of redemption, Transcendentalism, with Emerson, saw nature as "all beauty and commodity." Transcendentalism was a "systematic subjectivism" – a "sham system of Nature." Santayana was also to conclude John Dewey's humanistic pragmatism in the anthropocentric, genteel tradition. The problem, for Santayana, was that Western religion and philosophy were failing to provide any curbs or restraints on the developing urban–industrial society – the "American Will." If anything, they were providing a justification for the technological domination of nature. Santayana claimed that only one American writer, Walt Whitman, had fully escaped the genteel tradition and anthropocentrism by extending the democratic principle "to the animals, to inanimate nature, to the cosmos as a whole." Whitman was a pantheist; but his pantheism, unlike that of the Stoics and of Spinoza, was unintellectual, lazy, and self-indulgent." Santayana looked forward to a new non-anthropocentric revolution in philosophy – a "noble moral imagination" – of which Whitman was the beginning.[33]

Still alive in California at the time of Santayana's address was the one innovative American who most fully exemplified this non-anthropocentric "noble moral imagination." John Muir overcame his Calvinistic upbringing while studying science and Transcendentalism at the University of Wisconsin in the 1860s. Muir walked out of a career as a technological genius as a young man. He developed his nonanthropocentric philosophy while walking 1000 miles from Indiana to the Gulf of Mexico in 1867. For the next ten years, Muir wandered through Yosemite and the High Sierras developing ecological consciousness while arriving at the major generalizations of ecology through direct, intuitive experiencing of interrelatedness. Muir's development of ecological consciousness has been greatly underrated and misunderstood; while he has been referred to as mainly a "publicist" for the wilderness preservationist movement, he overcame the subjectivism of Transcendentalism to a much greater extent than did Thoreau. As Raymond Dasmann points out, those like Thoreau and Muir who advocated wilderness protection were proclaiming "the necessity of protecting at least some areas in which nature could remain intact against the destructive forces of civilization." Muir's battles at the turn of the century for Hetch Hetchy and with Pinchot over the status of the "forest reserves" (the "preservation–conservation" controversies) can now be seen as clashes between the biocentric egalitarianism of the emerging ecological consciousness and the old anthropocentric philosophy of Resource Conservation and the domination of Nature. Kevin Starr claims that "Muir became for Californians an avatar and prophet of all that the Sierra promised: simplicity, strength, joy, and affirmation . . . As a public figure he set a standard of what Californians should be . . . ."[34]

IV

The necessity for a new philosophy of nature, recognized by Muir and Santayana at the turn of the century, and revitalized during the 1960s and 1970s, can now no longer be ignored. The urban–industrial society is a dinosaur causing immense destruction in its death throes. New intellectual–social paradigms for postindustrial society are emerging. The paradigm which embodies contemporary ecological consciousness is called the "deep ecology movement." The Norwegian philosopher Arne Naess distinguished between "shallow and deep ecology" in 1972: "The Deep Ecology Movement gathers inspiration from the field ecologist (such as Rachel Carson) and is a rejection of the man-in-environment image in favor of the *relational, total-field image*. Organisms are knots in the biospherical net or field of intrinsic relations."[35] Biospherical egalitarianism "in principle" is asserted – "an awareness of the *equal right* of all things *to live and blossom* into their own unique forms of self-realization. Other principles are those of diversity, symbiosis, anti-class posture, local autonomy, and decentralization.[36] Ecological consciousness recognizes a spiritual *reciprocity* between humans, animals, and the land.[37] Inspiration for this movement is drawn from St. Francis, Spinoza, Muir, Jeffers, Aldo Leopold's ecosystem ethic (man is a "plain citizen" of the biotic community), Heidegger (we must "let beings be"), Taoism–Buddhism (humans *flow* with nature rather than trying to *control* it), the religions and ways of life of American Indians, and the science of ecological interrelatedness.[38] Theodore Roszak discussed most of the concerns of deep ecology in his brilliant book, *Where the Wasteland Ends* (1972); from the diminishment of consciousness and the spread of the "artificial environment" as the result of technological society, the threat of democracy from technocratic elites, the need for a resacrilized nature and sacramental perception, to the need for decentralization and for new democratic ecological/spiritual/anarchist communities.[39]

The ecopoet Gary Snyder is a major spokesman for deep ecology who provided an ecological utopian vision in "Four Changes" (1969): "If man is to remain on earth he must transform the five-millenia long urbanizing civilization tradition into a new ecologically-sensitive harmony-oriented wild-minded scientific/spiritual culture. What we envision is a planet on which the human population lives harmoniously and dynamically by employing a sophisticated and unobtrusive technology in a world environment which is *left natural.*[40]

A rival paradigm now emerging among many leading intellectuals, scientists, technologists, and business leaders is variously known as the New Age-Coevolution-Aquarian Conspiracy. The new spirituality, they proclaim, often based upon Catholic theologian Teilhard de Chardin, remains radically anthropocentric. Their conception of ecology largely remains that of Resource Conservation and Development. They still seem overly-infatuated with *obtrusive* technological gadgetry and "progress," the colonization of space and, most ominously, a human take-over of ecological systems and the evolutionary processes on this planet.[41]

Ecological consciousness in the person of Muir battled against inappropriate water development (Hetch Hetchy) and the conversion of forest ecosystems into managed "tree farms." The Age of Ecology, with Rachel Carson, Paul Ehrlich, and others, was a battle against pesticides, pollution, resource depletion, overpopulation, and overdevelopment. In the 1980s, the major issues appear to be the defense of the natural, global evolutionary process itself, and the protection of wild species and habitat against the onrushing "artificial environment" and man's complete domination of the planet, as we struggle toward decentralization and the establishment of human-sized ecological-spiritual communities.

NOTES

[1]See Huxley, "The Politics of Ecology," Center for the Study of Democratic Institutions, Santa Barbara, CA, 1964, and Huxley's ecological utopia, *Island* (New York: Harper & Row, 1962). A discussion of the development of Huxley's ecological consciousness occurs in Del Ivan Janik, "Environmental Consciousness in Modern Literature," in J. Donald Hughes & Robert Schultz (eds.) *Ecological Consciousness*, Univ. Press of America, 1981. Excellent contemporary statements of the problems facing society are Donald Higgins, *The Seventh Enemy* (New York, McGraw Hill, 1978); William Catton, Jr., *Overshoot: The Ecological Basis of Revolutionary Change*, Univ. of Indiana Press, 1980.

[2]For carefully documented summaries of the contemporary environmental crisis, see G. Tyler Miller, *Living in the Environment*, 2nd ed. (New York: Wadsworth, 1979); Paul and Anne Ehrlich, John Holdren, *Ecoscience* (San Francisco: W. H. Freeman, 1977).

[3]Commoner, *The Closing Circle* (New York: Knopf, 1971), p. 41. See also

William Murdoch & Joseph Connell, "All About Ecology," in Ian Barbour (ed.), *Western Man and Environmental Ethics* (Reading, Mass.: Addison-Wesley, 1973).

[4] John Passmore, *Man's Responsibility For Nature*, (New York: Charles Scribners Sons, 1974), chs. 1 & 2. See also Frederick Turner, *Beyond Geography: The Western Spirit Against The Wilderness* (New York: Viking, 1980).

[5] Rene Dubos, *The Wooing Of The Earth* (New York: Charles Scribners Sons, 1980).

[6] See Paul Shepard, "Ecology and Man," in P. Shepard & D. McKinley (eds.), *The Subversive Science* (Boston: Houghton Mifflin, 1969).

[7] Robert Reid, on the condor breeding program and tragedy, see Janice Phillip, "Saving the California Condor," *San Francisco Chronicle*, 29 June, 1980; "Condor Recovery: Hands-on or Hands Off?" *Not Man Apart* 10 (November) 1980; Robert Reid, "The Condor's Last Flight," *San Francisco Chronicle*, 15 February, 1981.

[9,10] Norman Myers, "How Shall We Choose Which Ones to Save?" *Not Man Apart*, Vol. 11, No. 2, February 1981.

[11] John Livingston, *The Fallacy of Wildlife Conservation*, McClelland and Steward, Ltd., 1981. For similar arguments on the inadequacy of treating the nonhuman world as "resources" for humans, see the ecologist, David Ehrenfeld, *The Arrogance Of Humanism* (Oxford: Oxford Univ. Press, 1978). See also Paul and Anne Ehrlich, *Extinction: The Causes And Consequences of The Disappearance of Species* (New York: Random House, 1981).

[12] John Passmore,

[13] For a discussion and critique of the Resource Conservation and Development position, see John Rodman, "Resource Conservation: Economics and After," Pitzer College, CA, 1977; Bill Devall, "Reformist Environmentalism," *Humboldt Journal of Social Relations*, Vol. 6, No. 2, 1979. For a critique of the technological "resource" development orientation of modern univerities, see Brian Martin, "Academics and the Environment: A Critique of the Australian National University Center for Resource and Environmental Studies," *The Ecologist*, Vol. 7, No. 6, 1977; Livingston & Mason, "Ecological Crisis and the Autonomy of Science in Capitalist Society," *Alternatives*, Vol. 8, No. 1, 1978.

[14] Devall, "Reformist Environmentalism." For an account of the plans of the U.S. Forest Service to apply management techniques to people visiting campgrounds and wilderness areas, see Joseph Sax, *Mountains Without Handrails* (Ann Arbor: Univ. of Michigan Press, 1980), pp. 98–101.

[15] Loren Eiseley, Forward to David Brower (ed.), *Not Man Apart: Lines From Robinson Jeffers* (San Francisco: Sierra Club Books, 1965). Jeffers' philosophical poetry embodies a powerful spiritual/ecological critique of modern technological society; see George Sessions, "Spinoza and Jeffers on Man in Nature," *Inquiry* (Oslo), Vol. 20, 1977; Del Janik, "Environmental Consciousness in Modern Literature."

[16] Roszak, *Person/Planet: The Creative Disintegration of Industrial Society*, (New York: Doubleday, 1978). See also Raymond F. Dasmann, "Conservation, Counter-culture, and Separate Realities," *Environmental Conservation*, Vol. 1, No. 2, 1974. And see chapter 13 ("Wilderness, Culture, and Counterculture") in Roderick Nash, *Wilderness and The American Mind*, Revised ed. (New Haven: Yale University Press, 1973).

[17] See Theodore Roszak, *The Making of A Counterculture* (New York: Doubleday, 1969). The bureaucratic management mentality of Resource Conservation and Development is described as "Consciousness II" in Charles Reich, *The Greening of America* (New York: Random House, 1970).

[18]Leo Marx, "American Institutions and Ecological Ideals," *Science*, Vol. 170, 1970, pp. 945–952. See, also, Marx, *The Machine In The Garden* (Oxford: Oxford Univ. Press, 1964). For more contemporary writing in this vein, see Mary Austin, *The Land of Little Rain* (New York: Houghton Mifflin, 1903; Frank Waters, *The Colorado* (New York: Rinehart, 1946; J. Frank Dobie, *The Voice of The Coyote* (Lincoln: Univ. of Nebraska Press, Bison Books, 1961); Edward Abbey, *Desert Solitaire* (New York: McGraw-Hill, 1968). For commentary, see John Alcorn, *The Nature Novel* (New York: Columbia Univ. Press, 1977); Paul Brooks, *Speaking For Nature* (New York: Houghlin Mifflin, 1980).

[19]Marx, "American Institutions."

[20]Rachel Carson, *Silent Spring* (New York: Houghton Mifflin, 1962), p. 261. For a biography of Miss Carson, see Paul Brooks, *The House of Life* (New York: Houghlin Mifflin, 1972). For the aftermath of the pesticide controversy, see Frank Graham, *Since Silent Spring* (New York: Houghlin Mifflin, 1970). Robert Van den Bosch, *The Pesticide Conspiracy* (New York: Doubleday & Co., 1978). For a critique of technology and environment, see Commoner, *The Closing Circle*; Langdon Winner, *Autonomous Technology* (Cambridge, Mass.: MIT Press, 1977).

[21]Stewart Udall,

[22]William Catton, Jr. & Riley Dunlap, "A New Ecological Paradigm for Post-Exuberant Sociology," *American Behavioral Scientist*, Vol. 24, No. 1, Sept/Oct. 1980. This issue of ABS is devoted to an ecological paradigm change for the social sciences. Victor Ferkiss, "Technology: The Hidden Variable in American Politics," *Review of Politics*, Vol. 24, No. 3, July 1980.

[23]Stuart Udall, *The Quiet Crisis* (1963).

[24]Stan Steiner, *The Vanishing White Man* (New York: Harper & Row, 1976), p. 24. For further discussion and defense of the wisdom of "primitive" peoples, see Paul Shepard, *The Tender Carnivore and The Sacred Game* (New York: Charles Scribners Sons, 1973); Turner, *Beyond Geography*; C. Vecsey & R. Venables, *American Indian Environments* (Syracuse, NY: Syracuse University Press, 1980). For a statement by a contemporary American Indian spokesman, see Russell Means, "Fighting Words on the Future of the Earth," *Mother Jones*, December 1980.

[25]Stan Steiner, *The Vanishing White Man* (New York: Harper & Row, 1976).

[26]Lynn White, Jr., "Historical Roots of our Ecologic Crisis," *Science*, Vol. 155, 1967 (reprinted in G. DeBell, *The Environmental Handbook* (New York: Friends of the Earth Book, 1970)). See also White, "Continuing the Conversation," in Barbour, *Western Man and Environmental Ethics*; Adolph Holl, *The Last Christian: St. Francis* (New York: Doubleday, 1980). For a scholarly account of Christian puritan attitudes toward wilderness and Indians, see Roderick Nash, *Wilderness and The American Mind*, chs. 1-3. See also John Opie, "Frontier History in Environmental Perspective," in J.O. Stefen, *The American West* (Norman, Okla.: Univ. of Oklahoma Press, 1979).

[27]For an excellent history of the development of ecology since the 17th century, see Donald Worster, *Nature's Economy* (San Francisco: Sierra Club Books, 1977). See also Anne Chisholm, *Philosophers of The Earth: Conversations With Ecologists*, E. P. Dutton, 1972.

[28]For a discussion of Marsh, see Stuart Udall, *The Quiet Crisis*, ch. 6, and Douglas Strong, *The Conservationists* (New York: Addison-Wesley, 1971), pp. 29–37.

[29]Russell, *A History of Western Philosophy* (New York: Simon & Schuster, 1945), pp. 494, 788–9, 827–8. For recent attempts to defend the ecological views

of Marx and Dewey, see H. L. Parsons, *Marx And Engels on Ecology* (New York: Greenwood Press, 1977); D. C. Lee, "On the Marxian View of the Relationship between Man and Nature," *Environmental Ethics Journal* Vol. 2, No. 1, 1980; George Clark, "Dewey and Environmental Problems," *Bicentennial Symposium of Philosophy,* CUNY Graduate Center, 1976; Roderick French, "Is Ecological Humanism a Contradiction in Terms? – The Humanities Under Attack," in Hughes and Schultz, *Ecological Consciousness.* For criticism of all humanism as inherently anthropocentric, see Ehrenfeld, *The Arrogance of Humanism.*

[30] Arthur Schopenhauer, *Religion,* trans. T. B. Saunders (New York: Allen & Unwin, 1890), p. 112.

[31] Hugo is cited in Passmore, *Man's Responsibility For Nature,* p. 3. For a recent defense of Schweitzer's ethics and a rejection of biocentric egalitarianism, see William Blackstone, "The Search for an Environmental Ethic," in T. Regan (ed.) *Matters Of Life And Death* (Philadelphia: Temple University Press, 1980). For an excellent critique of Schweitzer, see Paul Shepard, *Man In The Landscape* (New York: Alfred Knopf, 1967), ch. 6. For the animal liberation position, see Peter Singer, *Animal Liberation* (New York: Random House, 1975); Regan & Singer, *Animal Rights and Human Obligations* (Inglewood Cliffs, NJ: Prentice-Hall, 1976); Michael Fox, *Returning to Eden,* Viking, 1980; Tom Regan, "Animal Rights, Human Wrongs," *Environmental Ethics,* Vol. 2, No. 2, 1980. For critiques of animal liberation as failing to distinguish domestic from wild animals, and failing to take an ecological perspective, see John Rodman, "The Liberation of Nature?" *Inquiry* (Oslo) Vol. 20, Spring 1977; J. Baird Callicott, "Animal Liberation," *Environmental Ethics,* Vol. 2, No. 4, 1980; Peter Steinhart, "The Fossil Horse," *Audubon,* March 1981.

[32] George Santayana, "The Genteel Tradition in American Philosophy," in Santayana, *Winds of Doctrine* (Charles Scribners Sons, 1926), 186–215. There is an interesting discussion of Santayana's address in William Everson, *Archetype West: The Pacific Coast As A Literary Tradition* (Berkeley: Oyez Press, 1976), pp. 54–60.

[34] Muir's most anti-anthropocentric passages occur in the journal of his walk to the Gulf (published as *A Thousand Mile Walk to The Gulf* (New York: Houghlin Mifflin, 1916), pp. 13642. See also Herbert Smith, *John Muir* (Boston: Twayne Publishers, 1965), pp. 41–48; "The Philosophy of John Muir" in E. W. Teale, *The Wilderness World of John Muir* (New York: Houghlin Mifflin, 1954); Bill Devall, "John Muir as Deep Ecologist," *Environmental Review* (forthcoming). Michael Cohen of Southern Utah State College has just written a new scholarly biography of Muir which brings out his objective Taoist parallels. For a discussion of Muir's conflict with Pinchot, see Roderick Nash, *Wilderness and The American Mind,* pp. 122--40. Kevin Starr, *Americans and The California Dream* (Oxford: Oxford University Press, 1973), p. 186.

[35] Arne Naess, "The Shallow and the Deep, Long-Range Ecology Movements: A Summary," *Inquiry* (Oslo), Vol. 16, 1973, pp. 95–100. Further discussions of deep ecology appear in Bill Devall, "The Deep Ecology Movement," *Natural Resources Journal,* Vol. 20, Spring 1980, pp. 299–322; Devall, "Ecological Consciousness and Ecological Resisting," 1981 (pub. forthcoming); Alan Drengson, "Shifting Paradigms: From the Technocratic to the Person/Planetary," *Environmental Ethics,* Vol. 2, No. 3, 1980, pp. 221–240.

[36] Naess,

[37] For a discussion of "reciprocity" in Heidegger, and among the American Indians, see Dolores LaChapelle, "Systematic Thinking and Deep Ecology," and Calvin Martin, "The American Indian as Miscast Ecologist," both in Hughes &

Schultz, *Ecological Consciousness*. See also Paul Shepard, *Thinking Animals* (New York: Viking, 1978); Frank Waters, *The Man Who Killed The Deer* (1942) (New York: Pocket Books, 1970).

[38] For a further description and bibliography of St. Francis, Spinoza, Robinson Jeffers, Aldo Leopold, and Heidegger, and of their contributions to deep ecology theorizing, see George Sessions, "Shallow and Deep Ecology: A Review of the Philosophical Literature," in Hughes and Schultz, *Ecological Consciousness*.

[39] For an account of D. H. Lawrence's ecological consciousness and his awareness of the need for decentralization and for new scaled-down organic communities, together with his influence on Huxley, see Janik, "Environmental Consciousness in Modern Literature." There is an interesting discussion of medieval ecological utopias in Carolyn Merchant, *The Death of Nature* (New York: Harper & Row, 1980). A recent ecological utopia is Ernest Callenbach, *Ecotopia* (New York: Bantam Books, 1975). A very important analysis of philosophical anarchism as the basis for new ecological communities occurs in Val and Richard Routley, "Social Theories, Self Management, and Environmental Problems," in D. Mannison, M. McRobbie, and Richard Routley (eds.) (Canberra: Australian National University, 1980), pp. 217–332.

[40] "Four Changes" was published in Snyder, *Turtle Island* (San Francisco: New Directions, 1974). See also Snyder, *The Old Ways* (San Francisco: City Lights, 1977); Snyder, *The Real Work* (San Francisco: New Directions, 1980); L. Edwin Folsom, "Gary Snyder's Descent to Turtle Island: Searching for Fossil Love," *Western American Literature*, Vol. 15, 1980.

[41] For a description of the New Age-Aquarian Conspiracy, see Marilyn Ferguson, *The Aquarian Conspiracy* (New York: St. Martin's, 1980); J. E. Lovelock, GAIA (Oxford: Oxford Univ. Press, 1980). For a critique of this paradigm as anthropocentric and unecological, see Bill Devall, "New Age Ecology: A Critique," and Sessions, "Review of Bonifazi, *The Soul of the World*," *Environmental Ethics*, Vol. 3, 1981. For Loren Eisley's ecological utopian vision and critique of "outer space" as a solution, see "The Last Magician," in Eiseley, *The Invisible Pyramid* (New York: Charles Scribners Sons, 1970). For another critique of shallow anthropocentric spirituality, see Jacob Needleman, *A Sense of the Cosmos* (New York: Doubleday, 1975).

The Challenge of Disappearing Species

Norman Myers

Norman Myers has been an active ecologist throughout his busy life. He is a member of such organizations as the International Association of Ecology, the International Society of Tropical Ecology, the International Council of Environmental Law and the International Society of Social Economics. He is an on-going consultant, among many others, to the Survival Service Commission, the World Wildlife Fund, the Environmental Mediation International, and the World Future Society. Dr. Myers is currently working on a World Wildlife Fund-sponsored project on the Triage Strategy for Threatened Species. He has published over 200 articles concerning wildlife, Africa and conservation. His books include The Sinking Ark.

Ask a man in the street what he thinks of the problem of disappearing species, and he may well reply that it would be a pity if the tiger or the blue whale disappeared. But he may add that it would be no big deal, not as compared with crises of energy, population, food and pollution – the "real problems." In other words, he cares about disappearing species, but he cares about many other things more: he simply does not see it as a critical issue. If the tiger were to go extinct tonight, the sun would still come up tomorrow morning.*

In point of fact, by tomorrow morning we shall almost certainly have one less species on Planet Earth than we had this morning. It will not be a

*This essay is based on Dr. Myers' latest book *The Sinking Ark*, New York: Pergamon.

charismatic creature like the tiger. It could well be an obscure insect in the depths of some remote rainforest. It may even be a creature that nobody has ever heard of. But it will have gone. A unique form of life will have been driven from the face of the earth for ever.

Equally likely is that by the end of the century we shall have lost one million species, possibly many more. Except for the barest handful, they will have been eliminated through the hand of man.

EXTINCTION RATES

Animal forms that have been documented and recognized as under threat of extinction now amount to over 1000. These are creatures we hear much about – the tiger and the blue whale, the giant panda and the whooping crane, the organutan and the cheetah. Yet even though 1000 is a shockingly large number, this is only a fractionally small part of the problem. Far more important are those many species that have not even been identified by science, let alone classified as threatened. Among the plant kingdom, these could number 25,000 while among animals, notably insects, the total could run to hundreds of thousands.

Extinction of species has been a fact of life virtually right from the start of life on earth 3½ billion years ago. At least 90 percent of all species that have existed have disappeared. But almost all of them have gone under by virtue of natural processes. Only in the recent past, perhaps from around 50,000 years ago, has man exerted much influence. As a primitive hunter, man probably proved himself capable of eliminating species, albeit as a relatively rare occurrence. From the year A.D. 1600, however, he became able, through advancing technology, to over-hunt animals to extinction in just a few years, and to disrupt extensive environments just as rapidly. Between the years 1600 and 1900, man eliminated around seventy-five known species, almost all of them mammals and birds – virtually nothing has been established about how many reptiles, amphibians, fishes, invertebrates, and plants disappeared. Since 1900 man has eliminated around another seventy-five known species – again, almost all of them mammals and birds, with hardly anything known about how many other creatures have faded from the scene. The rate from the year 1600 to 1900, roughly one species every four years, and the rate during most of the present century about one species per year, are to be compared with a rate of possibly one per 1000 years during the "great dying" of the dinosaurs.

Since 1960, however, when growth in human numbers and human aspirations began to exert greater impact on natural environments, vast territories in several major regions of the world have become so modified as to be cleared of much of their main wildlife. The result is that the extinction rate has certainly soared, though the details mostly remain undocumented.

We face, then, the imminent elimination of a good share of the planetary spectrum of species that have shared the common earth-home with man for millenia, but are now to be denied living space during a phase of a mere few decades. This extinction spasm would amount to an irreversible loss of unique resources. Earth is currently afflicted with other forms of environmental degradation, but, from the standpoint of permanent despoliation of the planet, no other form is anywhere so significant as the fallout of species. When water bodies are fouled and the atmosphere is treated as a garbage can, we can always clean up the pollution. Species extinction is final. Moreover, the impoverishment of life on earth falls not only on present society, but on all generations to come.

In scores of ways, the impoverishment affects everyday living right now. All around the world, people increasingly consume food, take medicines, and employ industrial materials that owe their production to genetic resources and other start point materials of animals and plants. These pragmatic purposes served by species are numerous and growing. Given the needs of the future, species can be reckoned among society's most valuable raw materials. To consider the consequences of devastating a single biome, the tropical moist forests: elimination of these forests, with their exceptional concentrations of species, would undermine the prospects for modernized agriculture, with repercussions for the capacity of the world to feed itself. It could set back the campaign against cancer by years. Perhaps worst of all, it would eliminate one of our best bets for resolving the energy crisis: as technology develops ways to utilize the vast amounts of solar energy stored in tropical-forest plants each day, these forests could generate as much energy, in the form of methanol and other fuels, as almost half the world's energy consumption from all sources in 1970. Moreover, this energy source need never run dry like an oil well, since it can replenish itself in perpetuity.

Any reduction in the diversity of resources, including the earth's spectrum of species, narrows society's scope to respond to new problems and opportunities. To the extent that we cannot be certain what needs may arise in the future, it makes sense to keep our options open (provided that a strategy of that sort does not unduly conflict with other major purposes of society). This rationale for conservation applies to the planet's stock of species more than to virtually any other category of natural resources.

SPECIES CONSERVATION AND ECONOMIC ADVANCEMENT

There is another major dimension to the problem: the relationship between conservation of species and economic advancement for human communities.

We know that the prime threat to species lies with loss of habitat. Loss of habitat occurs mainly through economic exploitation of natural environments. Natural environments are exploited mainly to satisfy consumer demand for numerous products. The upshot is that species are now rarely driven extinct through the activities of a few persons with direct and deliberate intent to kill wild creatures. They are eliminated through the activities of many millions of people, who are unaware of the "spillover" consequences of their consumerist lifestyles.

This means that species depletion can occur through a diffuse and insidious process. An American is prohibited by law from shooting a snowy egret, but, by his consumerist lifestyle, he can stimulate others to drain a marsh (for croplands, industry, highways, housing) and thereby eliminate the food supply for a whole colony of egrets. A recent advertisement by a utility corporation in the United States asserted that "Something we do today will touch your life," implying that its activities were so far reaching that, whether the citizen was aware or not, his daily routine would be somehow affected by the corporation's multifaceted enterprise. In similar fashion, something the citizen does each day is likely to bear on the survival prospects of species. He may have no wanton or destructive intent toward wildlife. On the contrary, he may send off a regular donation to a conservation organization. But what he contributes with his right hand he may take away with his half-dozen left hands. His desire to be consumer as well as conservationist leads him into a Jekyll-and-Hyde role. Unwitting and unmalicious as his role might be, it becomes more significant and pervasive every day.

Equally important, the impact of a consumerist lifestyle is not confined to the home country of a fat-cat citizen. Increasingly the consequences extend to lands around the back of the earth. Rich-world communities of the temperate zones, containing one-fifth of earth's population, account for four-fifths of raw materials traded through international markets. Many of these materials derive from the tropical zone, which harbours around three-quarters of all species on earth. The extraction of these materials causes disturbance of natural environments. Thus affluent sectors of the global village are responsible – unknowingly for sure, but effectively nonetheless – for disruption of myriad species' habitats in lands far distant from their own. The connoisseur who seeks out a specialty import store in New York or Paris or Tokyo, with a view to purchasing some much-sought-after rosewood from Brazil, may be contributing to the destruction of the last forest habitat of an Amazon monkey. Few factors of the conservation scene are likely to grow so consequential in years ahead as this one of the economic-ecologic linkages among the global community.

True, citizens of tropical developing countries play their part in disruption of natural environments. It is in these countries that most of the projected expansion of human numbers will take place, 85 percent of the extra 2 billion people that are likely to be added to the present world

The Muriqui monkey of Brazil, one of the world's most seriously endangered primates.

population of 4 billion by the end of the century. Of at least as much consequence as the outburst in human numbers is the outburst in human aspirations, supported by expanding technology. It is the combination of these two factors that will precipitate a transformation of most natural environments throughout the tropics. Equally to the point, impoverished citizens of developing nations tend to have more pressing concerns than conservation of species. All too often, it is as much as they can do to stay alive themselves, let alone to keep wild creatures in being.

Plainly, a lot of difference exists between the consumerdom of the world's poor majority and that of the world's rich minority. For most citizens of developing countries, there is little doubt that more food available, through cultivation of virgin territories (including forests, grasslands, wetlands, etc.), would increase their levels of nutrition, just as more industrial products available would ease their struggle for existence in many ways. It is equally likely that the same cannot be said for citizens of the advanced world: Additional food or material goods do not necessarily lead to any advance in their quality of life. The demand for products of every kind on the part of the 1 billion citizens of affluent nations – the most consummate consumers the world has ever known, many making Croesus and Louis XIV look like paupers by comparison – contributes a disproportionate share to the disruption of natural environments around the earth.

For example, the depletion of tropical moist forests stems in part from market demand on the part of affluent nations for hardwoods and other specialist timbers from Southeast Asia, Amazonia, and West and Central Africa. In addition, the disruptive harvesting of tropical timber is often conducted by multinational corporations that supply the capital, technology, and skills, without which developing countries could not exploit their forest stocks at unsustainable rates. Such is the role of Georgia Pacific and Weyerhaeuser companies from the United States, Mitsubishi and Sumitomo from Japan, and Bruynzeel and Borregaard from Europe. Similarly, the forests of Central America are being felled to make way for artificial pasturelands to grow more beef. But the extra meat, instead of going into the stomachs of local citizens, makes its way to the United States, where it supplies the hamburger trade and other fast-food businesses. This foreign beef is cheaper than similar-grade beef from within the United States — and the American consumer, looking for a hamburger of best quality at cheapest price, is not aware of the spillover consequences of his actions. So whose hand is on the chain saw?

AN INTERDEPENDENT GLOBAL COMMUNITY

Looked at this way, the problem of declining tropical forests is intimately related to other major issues of an interdependent global community: food, population, energy, plus several other problems that confront society at large. It is difficult to make progress on one front without making progress on all the others at the same time. This aspect of the plight of tropical forests — the interrelatedness of problems — applies to the problem of disappearing species in general.

Similarly, the advanced-nation citizen can hardly support conservation of species while resisting better trade-and-aid relationships with developing nations. The decline of tropical forests could be slowed through a trade cartel of tropical-timber-exporting countries. If the countries in question could jack up the price of their hardwood exports, they could earn more foreign exchange from export of less timber. For importer countries of the developed world, the effect of this move would be a jump in the price of fine furniture, specialist paneling and other components of better housing. Would an affluent-world citizen respond with a cry of protest about inflation, or with a sigh of relief at improved prospects for tropical forests? For a Third World citizen, it is difficult to see how a conservationist can be concerned with the International Union for Conservation of Nature and Natural Resources without being equally concerned with the New International Economic Order.

A second example concerns paper pulp. There could soon be a shortage of paper pulp to match present shortages of fuel and food. The deficit could be made good through more intensive exploitation of North American forests or through more extensive exploitation of tropical forests — both of which alternatives might prompt outcries from environmental groups. A third alternative would be for developed-world citizens, who account for five-sixths of all paper pulp consumed worldwide, to make do with inadequate supplies, in which case the cost of newsprint would rise sharply. So perhaps a definition of a conservationist could be a person who applauds when he finds that his daily newspaper has once again gone up in price.

HOW FAR SHOULD WE GO TO SAVE SPECIES?

Just as the whooping crane is not worth more than a mere fraction of the United States' gross national product (GNP) to save it, so the preservation of species in all parts of the planet, and especially in tropical regions, needs to be considered wthin a comprehensive context of human well-being. Anthropocentric as this approach may appear, it reflects the way the world works: Few people would be willing to swap mankind, a single species, for fishkind with its thousands of species.

So the central issue is not "Let's save species, come what may." Rather we should ask whose needs are served by conservation of species and at what cost to whose opportunities for a better life in other ways. Instead of seeking to conserve species as an overriding objective, we should do as much as we can within a framework of trying to enhance long-term human welfare in all manner of directions.

As we have seen, people already make "choices" concerning species. Regrettably, they do not make deliberate choices after careful consideration of the alternatives. Rich and poor alike, they unconsciously contribute to the decline of the species, in dozens of ways each day. Not that they have malign intentions toward wild creatures. According to a 1976 Gallup Poll, most people would like to see more done to conserve wildlife and threatened species – 87 percent in the United States, 89 percent in Western Europe, 85 percent in Japan, 75 percent in Africa, and 94 percent in Latin America (though only 46 percent in crowded India). Subsistence communities of the developing world have limited scope to change the choice they implicitly make through their ways of making a living. Rich-world people, by contrast, have more room to maneuver, and could switch toward a stronger expression of their commitment in favor of species. Meantime, through their commitment to extreme consumerism, they in effect express the view that they can do without the orangutan and the cheetah and many other species – and their descendants, for all ages to come, can likewise do without them. In theory, they would like the orangutan and the cheetah to survive in the wild, but in practice they like many other things more. However unwittingly, that is the way they are making their choice right now.

Fortunately, affluent-world citizens still have plenty of scope to make a fresh choice. They may find it a difficult one. If they truly wish to allow living space for millions of species that existed on the planet before man got on to his hind legs, they will find that entails not only a soft-hearted feeling in support of wildlife, but a hard-nosed commitment to attempt new lifestyles. While they shed a tear over the demise of tropical moist forests with their array of species, they might go easy on the Kleenex.

WHAT ARE SPECIES GOOD FOR?

Aesthetic Argument

Aesthetically, species add to the diversity and texture of life's fabric on earth. All are complex and interesting, even the smallest microorganism. Look at a diatom under a microscope, and you will see that in form it is as beautiful as any antelope or butterfly.

In fact, many people in affluent nations spend appreciable sums enjoying the spectacle of wild creatures. Bird watchers in the United States alone spend over $500 million each year. A growing number of people from North America, Europe, and Japan annually roam the earth in search of wildlife, spending billions of dollars to catch a glimpse of a rhino in Africa, a tiger in Asia, or a tapir in South America.

True, no person can hope to enjoy the sight of more than the most trifling fraction of all species on earth. Generally speaking, however, he appears to get a kick out of the mere idea that he shares the earth with the orangutan, the cheetah, and creatures so obscure in the depths of Amazonia that nobody has discovered them. At least such is the case in advanced societies, where people spend much time and money on television programs, films, and books about wildlife.

In the case of certain species, of course, people cannot express their interest by going to see the creatures even if they have the funds available. They would find it hard to take a look at, for instance, a blue whale, not only because the creature is now very rare, but because it lives in the high seas. So if the blue whale were to disappear, would people actually feel as impoverished as at the thought that they will never see a dodo? In point of fact, it is unlikely that even the keenest conservationist loses much sleep over the dodo – just as the most ardent Greek scholar does not overly concern himself that he must make do with the seven Sophocles plays that have survived out of the dozens written. Yet many people might genuinely feel that their quality of life would lose something if the blue whale were denied life at all. In that case, they should be enabled to continue to enjoy their "option satisfaction" in the whale's survival.

Still more to the point, many people would almost certainly feel impoverished if they woke one morning in the year 2000 to find that they could no longer derive satisfaction from, say, one million species that had been eliminated within the previous 20 years.

Ethical Argument

As implied, the aesthetic argument quickly shades off into an ethical one. Briefly stated, this argument urges that all forms of life on earth have a right to exist. Conversely, humanity has no right to exterminate a species. To push the point a stage further, one could well ask whether humanity has the right to precipitate, through the elimination of large numbers of species, a fundamental and permanent shift in the course of evolution. While the aesthetic argument is virtually a prerogative of affluent people with leisure to think about such questions, the ethical concept of man's "stewardship" for earth's other creatures is inherent in many religious and cultural traditions around the world.

Yet even this argument needs to be looked at carefully. Are we so sure that we wish to preserve all forms of life for their own intrinsic worth? Would many people not be glad to see the end of the virus that causes the common cold? And would the same not apply to whatever organisms contribute to cancer? These are unique forms of life with just as much "right" to existence as the giraffe or the whooping crane. We have now reached a stage when the smallpox virus has been backed into a corner; no human being suffers from the disease, and the organism exists only in the laboratory. Should man, by conscious and rational decision, obliterate this manifestation of life's diversity?

Similarly, it is unrealistic to postulate that a species represents an absolute value and cannot be traded off against other value(s) on the grounds that a unique, irreplaceable entity is beyond measurement of "worth." Virtually no value is considered by society to be absolute. Not even human life qualifies. To be sure, a person views his own life as an untradeable asset. But as a member of the community, he does not view human life in general as anywhere near an absolute. The rate of road-accident deaths each year – 55,000 in the United States, 110,000 in Europe, and over one-quarter of a million worldwide – amounts to an appalling loss of life, yet it is apparently considered by society as an "acceptable" price for rapid transportation. One simple way to reduce road accidents would be to enforce the use of seatbelts, airbags, and lower speed limits. But the cost of these safeguards is apparently reckoned too high to match the amount of human life saved.

Presumably, only one value represents an absolute, and that is the survival of life on earth. To this absolute value, species makes an absolute contribution: When there are no more species left there will be no more life. But not all species make the same amount of contribution, since some are considered by ecologists to be more important for the workings of their ecosystems, by virtue of their numbers, biomass and energy flow. So the value of a species, far from being absolute, is very much a relative affair.

Especially difficult is the question of relative value between human life and other species. Already the conflict is plain to see in many localities, where people compete with wildlife for living space. This conflict is going to get worse, fast. True, in a few instances there is no clash; the blue whale encroaches on no human environment for its survival needs, and thus its demise would be all the more regrettable. For most species, however, the problem is basic: Sufficient habitat for them means less habitat available for human communities with their growing numbers.

How is this conflict to be resolved? The problem raises complex ethical issues that deserve in-depth treatment elsewhere. Suffice it to say here that many conservationists could probably accept the elimination of a species if it could be finally demonstrated that the creature's habitat would produce crops to keep huge communities of people alive (and provided that the food could not be grown elsewhere, that the people could not find any other means to sustain themselves, that the species could not be translocated to an alternative location, etc.). If the situation were reduced to bald terms of one species against one million people, it might well be viewed as a tolerable if regrettable tradeoff.

But the prospect facing humankind in the next few decades is a whole different ball game. To allow expanding numbers of people the amount of living space they seem to think they need, at least one-fifth (possibly one-third, conceivably one-half) of all species on earth may well be driven to extinction. Would this be in the best interests of human communities within the short-run future, let alone generations of the longer-run future? In view of the utilitarian benefits for agriculture, medicine, industrial materials that stem from species' genetic resources, it is virtually certain that humankind would suffer greatly through the disappearance of, say, one million species.

Such, then, is the nature and scale of the ethical conflict we now confront. It is a challenge that merits much more attention than it has hitherto received from conservationists, technologists, political leaders, economist planners, and whoever else determines the future course of our earth home.

Ecological Diversity and Stability

It is often argued that the more numerous an ecosystem's species, the greater is the ecosystem's stability. Much evidence supports and association between these two characteristics. For one thing, more species can use the sun's energy more efficiently than can a few. For another thing, an ecosystem can probably withstand perturbations better if each species can depend on many rather than few food sources, and be regulated by many rather than few predators – whereupon the eggs-in-one-basket effect is reduced.

Not that the idea is to be taken in the simple sense that variety is the essence of life. The relationship is far more complex. Diversity in this context refers to quality as well as quantity of differences among species, while stability can refer to numbers and relative abundance of species, or to dominance by a few species. So to assert that diversity equals stability is to overstate the case. A more concise way to express the situation might be to say that diversity and stability have had evolutionary relationships that run parallel without being causal. Alternately, we can say that high environmental stability leads to higher community stability, which in turn permits, though is not determined by, high diversity of species.

As many aspects of the challenge of conserving species, the problem of "how much diversity fosters how much stability?" boils down to a question of difference between marginal and absolute losses. If an ecological community with fifty species loses one, that may be considered, in certain circumstances, akin to a human being having a toe amputated – a bother, but no great setback. If three or five species disappear, that is like losing a foot – serious, but not necessarily critical to the foot-owner's survival. If twenty are eliminated, that is a basically different situation, akin to a human losing much of his digestive system through surgery. Leaving aside for a moment the fact that the loss of any species amounts to an irretrievable loss of a unique resource, and considering only the contribution of species to ecological stability, the loss of an individual species here or there, while regrettable, is hardly so catastrophic as it is sometimes represented. By contrast, if a significant segment of the species spectrum disappears, that is an altogether different order of loss, since it may strike at the adequate functioning of all life in the community at issue. So a key question arises: Where do losses shift from "regrettable but marginal" to "critical or worse"?

To tackle this issue at the global level, we can surely say, at one extreme, that the loss of one species out of a total of (say) 5 million is a pretty minor matter, while at the other extreme, the loss of 5 million species out of 5 million would amount to total catastrophe. Backing off one step from the latter situation, it is difficult to imagine a species that could exist entirely on its own on earth; and a community of a bare ten species, or even 100, or possibly 1000, might find themselves in a very precarious position. To revert to the other end of the gradient of possibilities, how many species can be lost before the impoverishment is no longer marginal in terms of stability for the planetary ecosystem? Or, to break the question down one stage further, how many species can be lost from regional ecosystems with scant consequence for the planetary ecosystem?

These are all problems of fundamental ecology that deserve urgent attention. Since we are clearly going to lose many hundreds-of-thousands of species before the end of the century, we need to know which ones we can "best afford" to lose, which ones would certainly leave major ecosystems with critical injury if they disappeared, which ones should be saved because their loss could precipitate ecological breakdown whose dimensions we can hardly start to envisage, and which ones should be preserved virtually at any cost. Conservation efforts need to be greatly expanded, but they need also to become more selective: The time is past when we can achieve much by running hither and yon with buckets of water.

A TRIAGE STRATEGY FOR SPECIES?

Even with a several-times increase in funding to assist threatened species, we could not save all those that appear doomed to disappear. The processes of habitat disruption are too solidly underway to be halted completely. Since man is intervening in the evolutionary process with all the impact of a major glaciation, he should do it with as much conscious awareness of what he is about as he can muster. That is to say, now that he is committed to playing God and determining the extinction of large numbers of species, he might as well do it as selectively as possible. But how to accomplish this? If we cannot be sure of the details, can we at least establish the right direction to move in?

These are large questions to ask. How are we to decide which species shall be allowed to go extinct through our deliberate decision and thereby concentrate our conservation efforts – limited as they are bound to be – on "more deserving" species? This would mean that certain species would simply disappear because we pulled the carpet out from under them. We might abandon the Mauritius kestrel to its all-but-inevitable fate and use the funds to proffer stronger support for any of the hundreds of threatened bird species that are more likely to survive. In short, a proportion of species would disappear through human design. Agonizing prospect though this might be, it is better than allowing species to disappear merely through human default.

An approach along these lines would amount to a "triage strategy" for species. The term derives from French medical practice in World War I, when battlefield doctors found there were more wounded than they could handle. So they assigned each soldier to one of three categories: first, those who could certainly be helped by medical attention; second, those who could probably survive without attention; and third, those who were likely to die no matter how much attention they received. The first category absorbed pretty well all the medical services available, so the other two categories were ignored. If a strategy along similar lines were applied to threatened species, it would amount to a more rational approach than that practiced hitherto. It would be systematic rather than haphazard, and it would enable conservationists to make best use of their finances and skills.

How would choices be made? How could we decide between the Bengal tiger and a crab in the Caribbean? Should we concentrate on remnant patches of rain forest in countries that have experienced decades of destruction, or should we try to "lock away" vast tracts of forest in regions that have been little touched? These will be difficult decisions. A start could be made through systematic analysis of what makes some species more susceptible to extinction than others; for example, sensitivity to habitat disruption, reproductive capacity, and "K-selection" traits. In addition to these bioecological factors, there is need to consider economic, political, legal, and sociocultural aspects of the problem: The Bengal tiger requires large amounts of living space in a part of the world that is crowded with human beings, but it could stimulate more public support for conservation of its ecosystem (and thereby help save many other species) than could a less-than-charismatic creature such as a crab. When we integrate all the various factors that tell for and against a species, we shall have a clearer idea of where best we can apply our conservation muscle.

Many tough decisions will have to be made. Nobody will like the challenge of deliberately consigning certain species to oblivion. However, insofar as man is certainly consigning huge numbers to oblivion, he might as well do it with as much selective discretion as he can muster. Having goofed at playing Noah, he might try to do a better job of playing God.

Current Development Efforts in the Amazon Basin

Emilio F. Moran

Emilio Moran: Associate Professor of Anthropology at Indiana University. His books include Human Adaptability: An Introduction to Ecological Anthropology*(1979), and* Developing the Amazon*(1982).*

INTRODUCTION

The decade of the 1970s was marked by a quickened pace in development activities in the Amazon Basin. The pace was set by Brazil, which in 1971 announced a Program of National Integration which included an ambitious program of highways cutting north-south and east to west across the Basin. The roads were accompanied by settlement schemes to populate the Amazon with rural peoples from throughout Brazil – "to integrate [the Amazon], so as not to give it up" (integrar para não entregar). Since then, the Radar da Amazônia (RADAM) Program has uncovered vast mineral deposits which have led to a shift in development emphasis – away from agropastoral and toward mining resource development (Moran 1982; Santos 1981). This review emphasizes the

Brazilian development efforts due to their magnitude and the proportion of the Basin under their influence.

Before discussing the most current development activities in the Amazon Basin, it would be useful to quickly review the characteristics of Amazonian development through time. During the colonial period, the Amazon provided exotic spices for Europe and labor for the coastal plantations. The extractive sector overshadowed the agricultural sector throughout the colonial period. Missionaries and lay people entering the Amazon tried to aggregate the Amerindians in villages where they could be christianized and their labor used to extract products of value for commerical trade. The expeditions to capture the Indians had a devastating effect. Disease and war decimated the populations by as much as 95%, according to some estimates (Denevan 1976; Hemming 1978). The difficulty in maintaining an adequate number of workers from this population, the transportation difficulties, and the presence of more valued products along the coast and in the highlands led to a declining interest in the Amazon beginning in the late 1600s.

The presence of mestizos throughout the Amazon maintained the presence of national society and the flow of goods and information between some native peoples and the outside. The mestizos in the interior of the Amazon developed distinct regional sub-cultures: characterized by Amerindian methods of subsistence and food preferences, while simultaneously following modes of dress, religion and social organization of Iberian origin (Wagley 1953; Moran 1974). Many of the native peoples fled deep into the forest, above the rapids that interrupt river transport along the fall line, and achieve a relative degree of isolation – both biological and cultural – from the outside.

The one event which disturbed this relative isolation was the Rubber Boom, lasting approximately from 1870 to 1920. During this period native peoples and immigrants were enlisted in the collection of rubber from the widely scattered trees through a system of debt peonage, which still exists in some isolated areas of the Amazon. The system charged exorbitant prices to the collector for basic products purchased and discounted these debits from the production of rubber. Cheating on weights was not uncommon, and the trader was the law on this riverine frontier (Collier 1968). The Rubber Era ended with the native peoples further depopulated by disease and exploitation.

The Second World War brought renewed efforts at exploiting natural rubber to provide this strategic product to the Allies, whose supplies from Malaysia were cut off by Japan. Old and new rubber barons enlisted peoples from the coast or highlands to extract rubber. Native peoples were also recruited. This short-lived boom began a process of internal colonization by nationals that was to steadily increase to the present time.

HIGHWAYS AND REGIONAL DEVELOPMENT

The end of World War II returned the Amazon economy to a stagnant condition, but the seeds of change had been planted. The major impetus for construction of highways into the upper Amazon took place in the Andean countries and came from the petroleum exploration activities of the Royal Dutch Shell Co. in the 1930s and 1940s. These roads of penetration were quickly populated by spontaneous migrants hoping to find work in prospecting operations and to claim land (Christ and Nissly 1973). This trend has intensified since that time. The Carretera Marginal de la Selva, begun in 1948, served not only to access the lowlands, but also "as a mesh to control the Brazilian bull" – as Brazil's neighbors put it, since Brazilian expansionism is feared throughout South America (Tambs 1974). Brazil has countered with the Transamazon Highway and a series of north-south highways that integrate the Basin to the rest of the country (Moran 1981).

Since 1964 Brazil has had the most coherent Amazonian development strategy. Their strategy is inspired by the geopolitical writings of Goldberry do Couto e Silva (1957). Since 1964 the Brazilian authorities have steadily announced a series of interventions: a) Operação Amazônia (Mahar 1979); b) replaced the bureaucratically bottlenecked and understaffed SPVEA with SUDAM (SPVEA 1960; Cavalcanti 1967); c) transformed the Banco da Borracha into a regional development bank (Banco da Amazônia); d) in 1971 announced a Program of National Integration; e) in 1972 announced a I Plan of National Development with an explicit Plan for Amazonian Development; f) in 1975 announced a Polamazonia Program simultaneously with a II Amazon Development Plan; g) in 1980 the President called for national reflection on how best to develop the region; and h) in 1980 announced a III National Development Plan in which Amazon development became the responsibility of the private, rather than the public, sector.

Amazonian development policy in the 1970s and 1980s has sought to overcome the region's underdevelopment through subsidization of investment, especially in large-scale projects. The overall effect of these policies has been to reduce the cost of land – already the least expensive production factor. At the same time, however, the cost of qualified managers has increased due to the historic scarcity in their supply and the inability of institutions to increase the supply relative to the rapid demand created by such development schemes (Norgaard 1979).

A cheap land policy, encouraged by subsidization of capital, and the absence of a tax policy to bring the cost of land use closer to the cost of other, and more dear, production factors has brought about severe environmental and social consequences (Norgaard 1979). Native peoples throughout the Amazon are being pushed off their lands, and serious conflicts have erupted between aborigenes and colonists (Davis 1977; Wagley 1977).

As the sections that follow will show, the greater constraint to Amazonian development is not the lack of land, labor or capital. Rather, the lack of management capacity at all levels of institutional functioning and the structural inability to use expertise at lower levels is the most severe problem. Given the absence of adequately developed human resources for development, the native peoples of the Amazon can play an important role in the adjustment of production to the realities of the Amazonian environment (Posey 1982). Amazonian native and peasant cultivators have an accumulated fund of knowledge that could provide solutions to the problems of coping with high rainfall, high temperatures, low soil fertility, and insect/pest invasion of cultivated fields. Unfortunately, very little is being done to protect this fund of human knowledge.

INDIANS AND THE LAND

The Indians are treated as wards of FUNAI (the Indian Protection Service). FUNAI has procceded slowly at the task of demarcating Indian lands due to the understaffed status of the organization and their susceptibility to pressures from the private sector. The inability of FUNAI to stop the Br-80 highway from cutting across the northeast corner of the Xingu Reservation is but a reflection of their incapacity to carry out the task assigned to them.

The problem is thought to be one of institutional focus. FUNAI seems fascinated with the process of "pacification" – but to neglect the care and resources that should follow the first contacts and the resettling of the Indians in protected land areas (Cultural Survival 1981). Without adequate provision for health care and for controlling access of frontiersmen to the Indians, the Indian is an easy prey to the common cold, influenza and measles. They continue to be decimated by disease at rates that probably approximate the earliest contacts in 1500. Simple vaccination programs are often undertaken only after infections have caused severe damage. The laws and the institutions already exist to protect the Brazilian Indians. What is needed is the politcal will to harness the human resources to implement the law and carry out the institutional priorities (Ibid).

The devastating effects of development projects upon Amazonian Indians can only be guessed at from the scattered facts that are available. In less than 20 months from the time the Cuiaba-Santarem cut through Kren-Akorore territory, their numbers declined from 400 to 70 people (Heelas 1978). The pacification of the Parakanã, during the Construction of the Transamazonian highway, meant the death of 45% of their population in less than a year (Bourne 1978). The Yanomamo are said to have suffered a loss of 22% of their population in four villages during the first year when the Perimetral Norte road was under construction (Ramos 1980).

If the Indian situation looks bleak in Brazil, it is even more chaotic in the other Latin American countries. Most other countries in Latin America do not have the equivalent of a FUNAI and also lack a consistent Indian policy. The emphasis of most countries is on "integration" and "development." In many ways the indigenous problem is more pressing in Ecuador, Peru, Colombia, and Bolivia than it is in Brazil. The numbers of indigenous people per unit of Amazonian territory is greater. On the other hand, pressure from colonists over the past two decades has put them ahead in their capacity to act as a united people. The Shuar of Ecuador are now part of a federation that acts as a united front to protect their people's rights. Other groups are following suit throughout the Amazon.

The rights of indigenous people to land have become mixed with disputed borders between countries. The disputed border between Peru and Ecuador has led to the incorporation of Indians into the military and to the restriction on free movement that had characterized earlier frontier policy. The frontier has become increasingly militarized and the native peoples have been treated as spies by all sides, given their independence and traditional freedom of movement. Recent reports from anthropologists are replete with accounts of colonists and military personnel taking advantage of the situation and confiscating Amerindians hard-earned wild game and other forest products on the grounds that they have improper documentation or that they are spies.

In all the countries the problem is not an absence of legislation to protect the Indians but the lack of implementation. The Church remains one of the few active forces to protect the Indians. However, it has become an increasing problem in itself in that the various denominations become competitive and enhance rather than overcome the interethnic friction that serves to weaken native peoples vs. their exploiters. The missionaries, moreover, have not taken as active a role in guaranteeing the Indians' rights to land as they might, given their own dependent status in getting access to the frontier (cf. Wagley 1977; Vickers 1981; Hvalkof and Aaby 1981).

A new President of the National Indian Foundation (FUNAI) has improved the outlook for several aboriginal land questions in Brazil. The new President, Coronel Paulo Moreira Leal, has indicated a commitment to resolve some of the most difficult Indian land rights problems. One of the first things that Coronel Leal did when he took over in October 1981 was to decree three new reserves for the Nambiquara Indians of the State of Mato Grosso. The three new reserves would cover a land area of

330,580 hectares. Unfortunately, this victory appears to have been temporarily reversed through pressure from interested agribusinesses in Mato Grosso and the reserves are now, again, "under study." In February 1982, the long fought efforts to legalize a Yanomamo Park in Northern Amazonia appears to have been approved.

COLONIZATION BY HOMESTEADERS

A major feature of the 1970s development process in the Amazon Basin was state-intervention in Amazon colonization and settlement (Scazzocchio 1980). Unlike the common experience of the lowlands wherein settlement is spontaneous, and sometimes even ahead of road construction, the 1970s were noted for major State-directed projects involving not only the construction of roads but colonist selection, provision of services, credit, and other inputs to the farm sector. Perhaps the greatest effort of the decade was that of the Transamazonian colonization projects. A complex system of agricultural villages (agrovilas), service centers (agropolis), and development poles (ruropolis) provided an interlinked system of settlements to promote "rural urbanism" and attempted to overcome the ancient constraints posed by rural isolation and government neglect. However, the projects proved to be too complex to be managed by the existing managerial capacity of institutions. The age-old constraints of poor road maintenance, inadequate health services, and inappropriate credit extension facilities reappeared as major constraints to the productivity of the sector (reviewed in Moran 1981 and 1982).

High rates of attrition are a common experience in all colonization projects. They were as high as 95% in the first three years in the Alto Beni, 60% in Santa Cruz, and down to 17-30% in the government-directed projects of the Transamazon highway. While it has become common to use this as a measure of success or failure, it is probably not a very telling measure. More important is to note how stable settlement becomes after the initial 3-5 year adjustment period, whether farmers are able to provide for their needs and to produce a marketable product, and how the incomes of the migrants compare with those from their peers in regions of origin. Stearman (1982) notes that the 262 Japanese families that came to the colonization region of San Juan in 1956 have developed cooperatives, have electrified their settlements, and even mechanized their agriculture in an area of 35,000 hectares in Bolivia. The Mennonites have also migrated since the 1950s and have relatively productive farms, although they resist assimilation much more than the Japanese. It has been rather well established that the quality of roads into a colonization area is the single most important predictor of the performance of small farmers (Crist and Nissly 1973; Moran 1979, 1981; Stearman 1982).

Without roads, farmers cannot get products to market and their intensive investment of labor is lost.

1975 the government had decided to abandon its financial investment in small farmers and to favor instead large-scale developers. In 1975 then President Geisel announced a Polamazonia program which sought to focus regional development in certain key areas – agriculture, mining, hydroelectric power and forestry – in certain pole areas wherein there was an identified comparative advantage in these products. By 1976 the government had begun to transfer a large portion of the task of regional development to the private sector. By 1979 INCRA had authorized 25 private colonization projects, some as large as 500,000 hectares, in northern Mato Grosso and Southern Pará (Schmink 1980).

The interest of the private sector in the development of the Amazon Basin did not begin in 1975 with the shift in government priorities. Rather, it preceded that shift as noted in the creation in 1968 of a society, the Associação dos Empresários Agropecuários da Amazônia (AEAA) made up mostly of Saõ Paulo businessmen with an interest in taking advantage of the new fiscal incentives made available to corporations through changes in the tax law (Pompermayer 1979). However, the cost of this large scale development to the taxpayer is great indeed. The fiscal incentives remove taxes from the Treasury and convert it into investment while the development agencies put up the other 75% of the cost of developing an area. To date, most of the cattle development projects remain unprofitable as producers of beef, but they have increased in value many times over through the simple ruse of removing the forest and transforming it into a low grade pastureland. These large projects do not absorb much labor, but cost the economic system an amount that has been estimated at 195 million cruzeiros per job (Ibid).

As the Amazon moves into the 1980s the most resonant theme in the Amazon interior has become that of "the Struggle for Land" (Foweraker 1981; Schmink 1982). When the 1970s began, it was hoped that by declaring most of the areas opened for settlement, as under the National Security Law, that all previous claims could be ignored and that a simplified and clear land titling procedure could be put in place. However, local and external pressures were placed on *cartórios*, so that new claims to land were held up in red tape for a long time, and the last terrestrial frontier has become a fierce battleground between Indians, squatters, migrants, and the would-be developers with large-scale schemes in mind. The major constraints appear to be the cheap price of land before clearing; the vastness of the region with the consequent difficulty to police it; and the institutional incapacity to deal with the land titling process. In the process many rural peoples are losing their rights to land claims, others are being shot in what has become a Wild West environment, while still others are giving up and becoming workers for the large-scale developers entering the area (Gall 1978). The struggle for land has become so heated that the National Security Council has created in one case the Araguaia-Tocantins Executive Land Group (GETAT) to police the streets, patrol the forest, and expedite land titling.

The roots of the problem caused by the entrance of large scale developers into the Amazon are evident when one notes how quickly the traditional structure of economic relations that is at the roots of the agrarian crisis in Latin America has become a part of the Amazonian experience. Whereas the small cultivator, with land less than 100 ha., cultivates 66 to 97% of the land owned, the owners of plots over 1,000 hectares cultivate barely 26% of their holdings and 14% and less in properties over 10,000 hectares. As much as 30 to 40% of the land is in the hands of operators with properties over 1,000 ha. in size. What we have throughout most of the Amazon is accumulation in the value of land as it lies unproductive or as it actually declines in agronomic potential through improper management. Development policies actually cheapen the price of undeveloped land, and by increasing demand create a scarcity in the already inadequate supply of managerial talent. The result is that inadequate management is actually promoted while vast tracts of land are devastated for speculative profits.

The end result of the 1970s decade of Amazonian development is mixed at best. The road construction programs, ambitious though they were, have been mostly completed. The Transamazonian highway, the Cuiaba-Santarem highway, and the Porto-Velho-Manaus highway, have been completed. The only major road planned in the early 1970s that has been temporarily abandoned has been the North Perimeter Rim road – due in part to the oil crisis, and in part to the greater frequency of encounters with uncontacted native peoples. The cost of roads turned out to be much higher than originally imagined – often three to five times the estimated cost (Moran 1981). Among the reasons for these unforeseen costs one must note the absence of data bases on the topographic characteristics of the Amazonian terrain (which turned out to be far more undulating than suspected); underbidding on the part of private contractors; the addition of military bases to the cost of road construction to assure the fulfillment of national security goals along the vast frontier; and unforeseen costs of road maintenance due to high rainfall and inadequate drainage provisions (Smith 1981).

PROJECTIONS FOR THE 1980s

The decade of the 1980s already manifests a new thrust in development efforts. Among the great changes will be the introduction of large amounts of hydroelectric power. The first major hydroelectric project to be completed will be at Tucuruí, on the Tocantins river, and will deliver a

40 megawatt (MW) capacity by 1985. This large amount of power is needed to take advantage of the vast mineral resources that have been uncovered in the Amazon (Radam 1974-80). Bauxite and iron ore are the two most immediately promising resources.

The iron ore deposits at Serra dos Carajás are estimated at 18 billion tons, representing the largest concentration of high grade iron ore (66% pure) in the world. In order to export and/or refine the ore the state-run enterprise Companhia Vale do Rio Doce is constructing a 890 kilometer railroad to São Luis do Maranhão. European and Japanese financial groups, the World Bank and Brazilian private funds are being used to cover the costs of the project (already reaching US $4.9 billion).

This project is overshadowed by the planned Greater Carajás project – a regional development concept which calls for mineral, agricultural and forestry development of almost the entire eastern Amazon Basin. Cost estimates hover around US $60 billion, easily the largest venture of its kind ever undertaken in Brazil. The Greater Carajás project would involve not only the exploitation of iron ore but also bauxite, copper, gold, manganese and nickel reserves. Plans include some areas to produce staple crops such as rice, corn, beans, sorghum, sugar and manioc. Forestry projects would include rubber, eucalyptus, and babaçú palm.

Each year brings new surprises in the resources found in the Basin. A mineral deposit rivaling the Carajás area has been discovered at Mapuera in 1982. The site covers approximately 175,000 km² in the area where Para, Amazonas and Roraima meet. Mapuera appears to have a 60,000 ton cassiterite deposit, with areas of gold and other minerals as well. The Minister of Mines and Energy has barred mining claims in order to prevent the kinds of abuses that have marked other recent finds. This project's development may be hampered by a lack of capital due to the massive investments in Carajás and Jarí in recent years. The area is also scarce in basic infrastructure and services.

The much-publicized Jarí Project of American multi-millionaire Daniel Ludwig was sold in 1982. After long negotiations the region, made up of up to 1.5 million hectares, was brought by a group of 27 Brazilian entrepreneurs for $280 million, payable over 35 years. Antunes, the leader of the new holding corporation, bought the Kaolin mine – the biggest money maker at present. Ludwig began to invest in Jarí in 1967 but was caught in a net of ecological miscalculations, friction in dealing with Brazilians in government and research institutes, and a poor sense of the place of Brazil in the international scene. As time passed, Brazilians saw in Jarí all the bad memories of traditional imperialism and sought to drive Ludwig out. The economic climate of the past few years cooperated in forcing Ludwig to seek Brazilian help – which they refused to provide in the past 3-4 years and forced Ludwig to put his property on the market.

The project had been hailed in the 1970s as an important experiment. Exotic trees were introduced to produce paper pulp (*Gmelina arborea* and *Pinus caribea*). Irrigated rice production of up to 8 tons per hectare was achieved in the floodplain with two crops per year. An integrated bauxite plant was barged all the way from Japan to exploit the rich ore

PHOTO: Eduardo Galvao

Jurura Indian, Xingu River, Mato Grosso, Brazil.

deposits. However, the inadequacy of resource surveys led to costly mistakes. An area of 50,000 hectares had to be abandoned due to soil compaction caused by bulldozer weight and scraping. *Gmelina's* initial growth response levelled off and was less satisfactory than better known pulp producers such as eucalyptus. In short, Ludwig committed the same error that flawed government-directed colonization in the 1970s.

In Peru and Ecuador the equivalent of the mineral boom began earlier through the exploitation of petroleum. In the Peruvian Amazon four companies currently operate, including Petroperu. Although the impact on the environment has not been assessed, it is believed to be significant (Camino 1982). Among the practices that are likely to cause environmental damage is the use of dynamite to open up streams to barges, the commercialization of hunting to feed the workers, and the oil spillage that occurs through careless drilling.

After long and careful studies, The World Bank has now signed an agreement with Brazil to loan $320 million for the first phase of an integrated rural development project in the northwest of the Amazon. Total cost for the project is estimated at 1.5 billion. The program will help pave the Br-364 highway linking Cuiabá and Porto Velho, and help regularize the entitlement of migrants to Rondônia and northern Mato Grosso. The area, known as Polonoroeste, includes the newly created state of Rondonia and western Mato Grosso. Close to one million people now live in this region and exert considerable pressure on the Indian population in the area. The area is believed to be gifted with good soils and the government is promoting permanent crops rather than annuals in order to protect the soils. The loan includes funding for research and for providing the Nambiquara with health services and land demarcation.

The development of agroforestry holds potential for the protection of the soil and the creation of sustainable systems. At Jarí there are 20,000 hectares planted in an association of Caribbean Pine (*Pinus caribea*) and guinea grass (*Panicum maximum*). The cattle graze in the area and gained weight at a rate of 50 kilograms per hectare per year, while reducing pine growth by only 5% from that of pure stands. At the Pirelli rubber plantation in Marituba, Pará, cattle graze a cover crop of kudzu (*Pueraria phaseloides*) underneath rubber trees (*Hevea* spp.) and gain up to 75 kilos per hectare per year (Hecht 1982).

PROSPECTS FOR AGRICULTURE IN THE 1980s

Recent advances in agronomic research do not deny that the most serious problem to crop production in the Amazon are nutritional deficiencies of many of the soils (Sanchez et al. 1982). Nine of every ten hectares of Amazonian soils are likely to be deficient in either nitrogen or phosphorus, or both. How severe these deficiencies turn out to be depends on the crops grown. Some crops with preadaptations to low nutrient levels and soil acidity (e.g., cowpeas and manioc) will give reasonably good yields in low N and P soils, whereas others will hardly develop. 79% of the soils are likely to have toxic levels of Aluminum although, again, some varieties within several crop species are tolerant of these levels of Aluminum. 60% of the soils are deficient in calcium, sulfur and magnesium. In short, the principal problems of Amazonian soils are chemical rather than physical. Most of the soils have relatively gentle slopes and favorable soil structures that permit rapid water percolation and reduce surface run-off. Less than 4% of the Amazon presents any lateritic formation hazard when exposed to air. They occur in flat, poorly drained areas subjected to repeated wetting and drying. The response of most cultivators to the problems of chemical constraints to crop production has been to rely on swidden cultivation to provide the chemical correction for the deficiencies present.

Numerous studies in recent years have documented the agronomic good sense of traditional forms of shifting cultivation. For populations with no significant amounts of capital, the regular shifting of fields maintains economic levels of output relative to input of labor, and the soil does not become degraded beyond repair. The long forest fallow permits the return of most of the forest biomass – to be used again later in the transformation of forest into cultivated fields. Unfortunately, growing pressure on the land has begun to cut short the length of fallow periods, the practice of intercropping, and the avoidance of steep slopes. With the shortening of the fallow period, the economic returns from farming these soils decreases and the return of secondary growth is sometimes impeded by the now compacted and impoverished soils.

In an effort to improve the production systems of the humid tropics in order to reduce the pressure on forested land and the maintenance of yields through time, steps have been taken in some circles to preserve the value of traditional practices. Agronomists have demonstrated that crop yields on soils cleared by traditional slash-and-burn methods were superior than those cleared by bulldozer (Seubert et al. 1977). Bulldozing of tropical forest leads to soil compaction, decreased water infiltration rate, topsoil displacement and consequent loss of organic matter accumulated by litterfall, and loss of the nutrients in the ash with their beneficial raising of pH and nutrient levels in the soil. While there is no question than slash-and-burn is agronomically and economically sensible it has not been fully established how to keep these fields productive after one or two years of cropping – by which time the nutrients deposited in the ash have been absorbed by plants or leached by the elements.

What recent studies, such as those carried out for the past 8 years at Yurimaguas, Perú, suggest is that caution and continuity of experimentation are necessary before any given area can be protected as well as productive (NCSU 1975, 76-77, 78-79).

A number of management findings, however, can be generally applied: a) Hand clearing and burning of forest provides better initial production results at lower cost; b) Once cleared, priority should be given to establishing a solid canopy to protect the soil from erosion and leaching; c) Crop associations wherein nitrogen fixing and other forms of biological conservation need to be promoted; d) Soil nutrient levels should be monitored once or twice yearly to prevent nutrient imbalance leading to drop in yields; e) Experimentation with crops should begin with those that are preadapted to local conditions, especially in areas of acidic, low nutrient levels with al toxicity levels. Small areas of crops more sensitive to these chemical constraints should be planted but not to the neglect of guaranteed crops; f) Agronomic research should constantly be measured against the realities of the cultivators of the area in question and adjusted for fit to the laborland and capital constraints.

An alternative to cropping and pastures that has been proposed for Amazonia is the development of agroforestry. The seriousness of this alternative to current policy-makers is evident in the rapid development of cacao, coffee, and rubber tree crops in the colonization area of Rondonia (IBRD 1981). Agroforestry offers some obvious advantages towards the achievement of sustainable agroecosystems for the humid tropics. It usually assumes multiple canopies with more than one harvestable stratum – a strategy that helps to protect the soil, and takes advantage of the deeper rooting structure of trees in capturing nutrient inputs.

The simplest form of agroforestry is the introduction of tree crops into the secondary successional process. Native Amazonians appear to have done so regularly as a way of having a food supply when they traveled through the forest in search of game and other products (Posey 1982). More commercial in nature has been the practice of *taungya* in which commercial timber species are planted into the swidden plots and are allowed to develop during the forest fallow period. The commercial species are then harvested before the forest is prepared for burning. In Latin America it has been common to mix cash crops such as cacao and coffee with tree species as a way of getting two commercial harvests. Trees such as *Erythrina* and *Glircidia* that tolerate considerable pruning are often used to shade these crops (Hecht 1982).

CONCLUSIONS

The contemporary thrust in Amazonian development that began in the 1950s has changed our views of the region. The Amazon is no longer an Eldorado, nor expected to be a breadbasket. Nor do we really believe that it is a green hell or totally wild. We might have expected the experience of the 1950s and 1960s to have taught everyone that development planning is hard – and that implementation is even harder. The experience of the 1970s in Amazonian development reminds us once again that unless the bottleneck of inadequate managerial capacity is dealt with, the achievement of a sustainable development will forever elude us. Better knowledge of the Amazonian environment makes it possible for us to make management discriminations unimagined a decade before. The agronomic research at Yurimaguas demonstrates the potential of agriculture in the Amazon but the intensive input of management required to implement it. The findings from colonization studies show the incapacity of the State in administering settlements, and the significance of roads, health care, and credit in the performance of homesteaders. The current emphasis on hydroelectric and mining technologies does not address the fundamental problems of Amazonian development which are, as we have seen, problems of adequate environmental knowledge, training of personnel for implementation, and misplaced priority given to growth rather than human resource development.

REFERENCES

Anderson, Robin. 1976. "Following Curupira: Colonization and Migration in Pará (1758-1930)." Ph.D. Dissertation, Univ. of California Davis, Dept. of History.

Bourne, Richard. 1978. *Assault on the Amazon.* London: Gollancz.

Camino, Alejandro. "La Amazonia Peruana, Sus Recursos Naturales y La Ideología Political de Desarrollo Regional." Paper Presented at the Conference on Frontier Expansion in Amazonia. Center for Latin American Studies, Univ. of Florida, Feb. 3-8, 1982.

Cavalcanti, M. de Barros. 1967. *Da SPVEA à SUDAM.* Belém: Univ. Federal do Pará.

Centro Internacional de Agricultura Tropical (CIAT). 1982. "Amazonian Agricultural Land Use: Proceedings from a Conference." Cali, Colombia: CIAT.

Collier, Richard. 1968. *The River that God Forgot.* New York: Dutton.

Couto e Silva, Goldberry do. 1957. *Aspectos Geopolíticos do Brasil.* Rio de Janerio: Bibliotéca do Exêrcito.

Crist, Raymond and C. Nissly. 1973. *East from the Andes.* Gainesville: Univ. of Florida Press.

Cultural Survival. 1981. "In the Path of Polonoroeste: Endangered Peoples of Western Brazil." Cambridge, Mass.: Cultural Survival, Occasional Paper #6.

Davis, Shelton. 1977. *Victims of the Miracle.* New York: Cambridge Univ. Press.

Denevan, William, ed. 1976. *The Native Population of the Americas in 1492.* Madison: Univ. of Wisconsin Press.

Foweraker, Joe. 1981. *The Struggle for Land.* London: Cambridge Univ. Press.

Gall, Norman. 1978. *Letter from Rondonia.* Hanover, NH: American Universities Field Staff Reports, Nos. 9-13 (South America).

Hecht, Susanna. 1982. "Agroforestry in the Amazon Basin." In S. Hecht and G. Nores eds. *Land Use and Agricultural Research in the Amazon Basin.* Cali, Colombia: CIAT.

Hemming, John. 1978. *Red Gold: The Conquest of the Brazilian Indians.* Cambridge, Mass: Harvard Univ. Press.

Havalkof, Søren and P. Aaby, eds. 1981. *Is God an American? An Anthropological Perspective on the Missionary Work of the Summer Inst. of Linguistics.* Copenhagen: Internat. Work Group for Indigenous Affairs.

World Bank IBRD. 1981. *Brazil: Integrated Development of the Northwest Frontier.* Washington, DC: World Bank.

Mahar, Dennis. 1979. *Frontier Development Policy in Brazil: A Study of Amazonia.* New York: Praeger.

Moran, Emilio. 1981. *Developing the Amazon.* Bloomington: Indiana Univ. Press.

———. *Criteria for Choosing Successful Homesteaders in Brazil.* Research in Economic Anthropology 2 (1979):339-59.

———. 1974. "The Adaptive System of the Amazonian Caboclo." In C. Wagley, ed. *Man in the Amazon.* Gainesville: Univ. of Florida Press.

———., ed. 1982. *The Dilemma of Amazonian Development.* Boulder, Colo: Westview Press.

Norgaard, Richard. 1979. "The Economics of Agricultural Technology and Environmental Transformation in the Amazon." Paper prepared for a collaborative study between the Min. de Agricultura and CEDEPLAR/Univ. Federal de Minas Gerais, Brazil.

North Carolina State University. "Research on Soils of the Tropics." Annual reports. Raleigh, NC: Soil Science Dept., NCSU, 1975-79.

Pompermeyer, Malori. 1979. "The State and the Frontier in Brazil: A Case Study of the Amazon." Ph.D. Dissertation, Stanford Univ.

Posey, Darrell. 1982. "Indigenous Ecological Knowledge and Development of the Amazon." In E.F. Moran, ed. *The Dilemma of Amazonian Development.* Boulder, Colo.: Westview Press.

RADAM. "Levantamento de Recursos." Naturais. Vols. 1-9. Rio de Janeiro: Ministŕio de Minas e Energia, 1974-79.

Ramos, Alcida. "Development, Integration and the Ethnic Integrity of Brazilian Indians." In F. Scazzocchio, ed. *Land, People, and Planning in Contemporary Amazonia.* Cambridge, UK: Centre for Latin American Studies Occasional Publ. #3.

Sanchez, Pedro et al. "Amazon Basin Soils: Management for Continuous Crop Production." Science 216 (1982): 821-827.

Santos, Breno A. dos. 1981. *Amazônia: Potencial Mineral e Perspectivas de Desenvolvimento.* São Paulo: Editora da Univ. de São Paulo.

Scazzocchio, Francoise, ed. *Land, People, and Planning in Contemporary Amazonia.* Cambridge, UK: Centre for Latin American Studies Occasional Publ. #3, 1980.

Schmink, Marianne. 1977. "Frontier Expansion and Land Conflicts in the Brazilian Amazon." Paper presented at meeting of American Anthropol. Assoc.

———. "Land Conflicts in Amazonia." American Ethnologist 9 (1982): 341-57.

Seubert, C. et al. "Effects of Land Clearing Methods on Soil Properties and Crop Performance on an Ultisol of the Amazon Jungle of Peru." Tropical Agriculture 54 (1977): 307-21.

Smith, Nigel. "Colonization: Lessons from a Tropical Forest." Science 214: 755-761.

SPVEA. 1960. *Political de Desenvolvimento da Amazonia.* Belem: SPVEA.
Stearman, Allyn. 1982. "From Forest to Pasture." In E.F. Moran, ed. *The Dilemma of Amazonian Development.* Boulder, Colo.: Westview Press.
Tambs, Lewis. 1974. "Geopolitics of the Amazon." In C. Wagley, ed. *Man in the Amazon.* Gainesville: Univ. of Florida Press.
Vickers, William. "Perspectives on Indian Policy in Amazonian Ecuador." Paper presented at Conference on Frontier Expansion in Amazonia. Center for Latin American Studies, Univ. of Florida, Feb. 3-8, 1982.
Wagley, Charles. 1953. *Amazon Town.* New York: Macmillan.
– – –. 1977. *Welcome of Tears.* New York: Oxford.

On the Dire Destiny of Human Lemmings

William R. Catton, Jr.

William Catton, Jr. is Professor of Sociology at Washington State University. He is the author of From Animistic to Naturalistic Sociology *(1966) and* Overshoot: The Ecological Basis of Revolutionary Change *(1980). In addition to holding other elective offices in several professional associations, Dr. Catton was the first chairman of the Section on Environmental Sociology of the American Sociological Association. Dr. Catton has written or co-authored more than eighty published articles.*

OUR COLOSSAL PREDICAMENT

By some odd mutation afflicting mankind, what if children grew to twice the adult size attained by their parents? Further, what if the effect of this mutation were somehow cumulative, so that grandchildren grew four times as large as their grandparents, great-grandchildren eight times as large, and so on? With each successive generation requiring twice as much food and fiber per capita as the preceding generation to sustain life and comfort, then even without becoming any more numerous, latter-day giants would press much harder than their forebears upon the world's finite carrying capacity.

It is a fact of life, to be ignored at our deepening peril, that there is indeed some limit upon the amount of use of a given kind that a particular environment can endure year after year without degradation of its suitability for that use (Allen, 1967: 1). That is the meaning of "carrying capacity." Overuse can *reduce* an environment's suitability for a given sort of use (e.g. to support human life, or to sustain a particular form of human culture) even if the environment in question is an entire planet, such as the earth. Carrying capacity, though variable and not easily or always measurable, must be taken into account to understand the human predicament. A longstanding lack of public comprehension of this concept has brought on the basic malady of the human condition today – diachronic competition – for contemporary well-being is now obtained at the expense of our descendants (Catton, 1976: 262). Satisfying today's human aspirations requires massive deprivation of resources for posterity because of our sheer numbers and the state of our technological development, and because we have been oblivious to the way breakthroughs in the past that achieved lasting increments of human carrying capacity differ fundamentally from modern techniques that yield inherently temporary supplements to carrying capacity.

Underlying this predicament is a kind of "mutation" that has made each generation effectively more gigantic than its precedessors – a cultural transformation of the relationship between our species and the world's ecosystems. Therefore, conventional political nostrums will fail to remedy the problems arising from the situation we now face, and so will revolutionary agitation. We might sense the futility of merely political or even millenarian responses to our predicament if we ceased to call ourselves *Homo sapiens* and began to think of ourselves as *Homo colossus,* a species prosthetically equipped by cultural evolution to magnify with technology the per capita requirements and the impact of each passing generation upon its surroundings. If we came to think of a human being not just as a naked ape or a fallen angel but as a man–tool system, we might recognize how progress could become a disease. The more colossal our tool kit, the larger our resource appetites are and the more destructive toward our children's future become the routine activities of daily life.

According to the United Nations Statistical Yearbook, by 1971 the rate of energy consumption per capita in the United States was about sixty times greater than in India. Americans are thus more mobile than the people of India – which is to say, more space consuming. We do more things; we make much larger and more varied demands on the ecosystems we exploit than do Indians. We can interact with more people in more ways, so we have more ways of interfering with each other than do most Indians, although less competition for scarce food between living Americans than between contemporary Indians. Each colossal American, with his phenomenally high rate of energy use, has in effect sixty times as many "slaves" working for him as does each person in India. So it would be not too inaccurate to suggest that, for a realistic comparison of

American pressure upon carrying capacity with that of India, the ratio betwen these two nations' respective population densities should be multiplied by a factor of sixty. The actual number of American people per square kilometer of U.S. territory yields a density of about one-eighth that in India. But if each American is, by virtue of the technology at his direct or indirect command, about sixty times as colossal as each Indian, then American population *pressure* is equivalent to there being *twelve billion* people living the Indian pace of life within the United States. Because of our persistent adherence to obsolete ways of thinking about people, equipment, lifestyles, and environments, Americans do not commonly recognize that the United States is more overloaded than "teeming India" (Borgstrom, 1969; Day and Day, 1964).

We have mistaken temporary and unreliable extensions of carrying capacity for permanent expansion of opportunity and durable progress. To see into our future realistically, basic principles of ecology have become as essential as literacy, even though unique characteristics of the human species have been misinterpreted and have provided an excuse for supposing that ecological principles do not apply to mankind (Catton and Dunlap, 1980). We embarked upon a tragically hazardous course when (understandably but invalidly) we claimed independence from nature. Now, by various tactics of mental evasion, we try to believe extravagant dreams can yet be fulfilled, while many of the world's people are responding to ecological pressure in ways that dehumanize. Intensified competition for diminishing resources undermines decency and hope. Typical responses to pressure aggravate that pressure.

To keep from gravitating toward genocidal conflict, we must stop demanding perpetual progress. For quite nonpolitical reasons, governments and politicians cannot achieve the paradise they habitually promise. Political leaders who continue to dangle before their constituents enticing carrots that are becoming unattainable hasten the erosion of faith in political processes. Circumstances have ceased to be what they were when the once-New World's myth of limitlessness made sense.

In today's world we urgently need to see human society as inextricably part of a global biotic community in which human dominance has had, and is having, self-destructive consequences. There are already more human beings alive than the world's *renewable* resources can perpetually support. Modern societies therefore depend on rapid use of *exhaustible* resources. Thus, depletion of resources we don't know how to do without is reducing this finite planet's carrying capacity for our species.

Carrying capacity is also reduced by the accumulation of harmful substances unavoidably created by our life processes. There are so many of us, using so much technology, that these substances accumulate faster than the global ecosystem can reprocess them. Moreover, by overloading nature's reprocessing systems we are breaking down their already limited capacity for setting things right for us (Wurster, 1968; 1969).

TRAGIC ORIGINS OF MAN'S FUTURE

Viewed ecologically, the human saga is a tragic success story. It is the story of a world that was again and again "filled up" with human inhabitants only to have its limits raised by human ingenuity. A series of technological breakthroughs that took almost two million years enabled human populations repeatedly to take over for human use portions of the earth's total life-supporting capacity that previously had supported other species. The most recent breakthroughs, however, have had spectacular results but have committed us to drawing down finite reservoirs of materials that do not replace themselves within any human span of time.

In the beginning, some two million years ago, a species of human creatures had evolved from prehuman ancestors and occupied a niche in the web of life somewhat different from that of their prehuman ancestors. These first humans had acquired the use of fire – to warm themselves, to ward off predators, to cook and thus render digestible organic substances that otherwise would not have been usable by their bodies as nutrition. As newly human beings they could make and use simple tools and could teach their progeny these advantageous ways. Yet even with fire, tools, and traditions, early humans remained consumers of naturally available foodstuffs obtained from wild sources by hunting and gathering. So they could never have been very numerous. (Any one of our world's forty or so most populous cities today contains more people than inhabited the entire planet in those days.) Even so, these early humans were successful because they survived, reproduced, adapted, and continued evolving. Selection pressures imposed by their habitats evoked biological and cultural responses by perhaps 80,000 generations of human hunters and gatherers and gave rise to a descendant population with basically the inheritable physical traits we see among men and women today. Thus, by about 35,000 B.C. the humans on Earth were of our own species, *Homo sapiens*. There were probably about three million of them (Boughey, 1975: 251) still living by hunting and gathering. From the long time it had taken to attain even that number, we can be reasonably sure that it was close to this planet's carrying capacity for that kind of creature living in that way.

But the gradually evolving cultures of *Homo sapiens* eventually increased the earth's human carrying capacity by changing man's ecological role. About 35,000 B.C., someone discovered how much harder and farther a spear could be thrown if the thrower effectively lengthened his arm by fitting the end of the spear into a socket at the end of a hand-held stick. Someone else invented a way of propelling arrows – miniature spears – both faster and in a manner that permitted more accurate, line-of-sight aiming. The survival value of spear throwers (the *atlatl* or the *woomera*) and of bows, lay in the greater effectiveness those tools gave this hunter species. More of the world's game animals became nourishment for human bodies. The total human biomass on Earth more than doubled in a little over 1000 generations; between 35,000 and 8000 B.C., world population increased from about three million to about eight million (Coale, 1974: 43). Still, the average enlargement of numbers *per generation* was less than one-tenth of one percent.

But eventually there came another major breakthrough and another enlargement of Earth's human carrying capacity. Somewhere, women who gathered wild seeds for grinding into flour may have noticed how seeds they happened to spill on moist soil near where the family carried on its domestic activities might grow into plants and grow at least as well as in the wild. Development of this discovery into techniques of deliberate plant cultivation wrought a major transformation of our species' relation to nature's web of life: Humans began to obtain nourishment from a humanly managed portion of the biotic community.

This "horticultural revolution" turned some hunters and gatherers into farmers. It launched another increase in Earth's human population. It also made it possible for a miniscule but increasing fraction of any tribe to devote its time to activities not connected with obtaining sustenance. Social organization thus began developing along more elaborate lines than before, and the rate of cultural innovation could further accelerate. Each increment of technology gave mankind a competitive edge in interspecific competition, and this eventually enabled our own species to become the dominant member of ecosystems all over the world.

Around 4000 B.C., stone and bone tools were augmented and then superseded by metal tools. Bronze Age metallurgy further enhanced the ability of *Homo sapiens* to harvest nature's products rather than leaving them to be used by other consumer species. Metallurgy also stimulated a division of labor among increasingly specialized occupations; this cultural counterpart of speciation (within what nevertheless remained biologically a single species) was henceforth increasingly important in human dominance over the environment.

Around 3000 B.C., the plan made possible an early version of mass production. With this device, farmers began using nonhuman energy to turn the soil – energy supplied by the muscles of an ox or a horse. One farmer could now manage a larger piece of the ecosystem. But agriculture that used draft animals had to use some land to raise crops to be eaten by animals other than man. Because a farmer with a plow and a draft animal could farm enough land to feed himself, the animal, his own family – and perhaps have a bit to spare – some small but gradually increasing fraction of the population could now do other things besides raising food; and human groups could opt for further elaboration of their lives instead of simply expanding their numbers. When iron tools came into use, some 2000 years after the first plow, again some of the carrying capacity increment they afforded enhanced living standards for at least some groups.

The cumulative effect of these breakthroughs was to expand the world's human stock from about 8 million just before the horticultural revolution to about 350 million by around 1500 A.D. (Petterson, 1960: 872). The change, however, would have remained almost unnoticeable to those living through it, for the average increment per generation was still just over 2 percent. In effect, the starting population of 8 million was multiplied in one generation by a factor of 1.0212, and then that product was again multiplied in the next generation by 1.0212, and so was that product, and so on. The "interest" of 2.12 percent on the initial demographic "investment" was compounded each generation—compounded 380 times between 8000 B.C. and 1500 A.D.: $8{,}000{,}000 \times 1.0212^{380}$ = approximately 350,000,000. Even at low percentage rates of increase per generation, the "compound interest" pattern can produce great change when enough generations elapse (i.e., when the exponent is large).

Then came a different kind of breakthrough. Those human tribes, which had benefited from all of these past breakthroughs, were pressing hard on the limited, though repeatedly enlarged, carrying capacity of what was soon to be known as the Old World. But the tools and the knowledge available to these culturally most advanced groups enabled (and caused) some men to leave the land and seek their fortunes at sea. Less than two centuries after acquiring firearms, Europeans discovered lands they had not previously known existed. Their superiority in weapons enabled them to take possession of a New World that seemed almost empty to them because its prior inhabitants were still living by Stone Age methods at a hunter-gatherer or early horticultural level.

When Columbus set sail there were just under 10 hectares of Europe per European. Life was a struggle to make the most of resources that were no longer sufficient nor reliable. After Columbus serendipitously found a second hemisphere, and after monarchs and entrepreneurs began to make its lands available for European exploitation and colonization, the expanded European habitat had a total of 48 hectares of land per person—about five times the pre-Columbian figure (Webb, 1952: 17–18). This sudden and impressive increment of exploitable carrying capacity spawned new beliefs, new human relationships, new behavior, and new institutions. An aura of limitless opportunity evolved; one result was further acceleration of the growth of population. Between 1500 and 1800 A.D., the world's human load grew from 350 million to about 969 million (the midpoint in a range of estimates compiled by Durand [1967]). There had never been so huge an increase in so short a time. The average increase per generation during these three centuries was 8.86 percent, a rate probably rapid enough to be noticed by those among whom it was happening. Other social effects of the apparently infinite carrying capacity increment made growth seem almost synonymous with progress – tragically so.

THE TAKEOVER METHOD VS. THE DRAW-DOWN METHOD

The Europeans who began taking over the New World in the sixteenth and seventeenth centuries were not ecologists, of course, so they were not prepared to recognize that, in an ecological sense, this second hemisphere was already "full." Just as the planet could not support as many people when they all lived as hunters and gatherers as it could when some had advanced to the agrarian level, so a continent that was ecologically fully occupied by hunters and gatherers was bound to seem almost empty to invaders from lands exploited by agrarian activity and accustomed to their greater density of settlement. All over the world, therefore, Europeans after 1500 acted on the premise that it was only fair and reasonable to "put to good use" those "unused" or "underused" lands – which were being used as fully as possible by nonagrarian non-Europeans. Native peoples in the New World were relegated to "reservations" not extensive enough to support their previous numbers in their traditional ways.

But enlarging carrying capacity for one's own kind by invading and usurping lands already occupied by others was essentially what mankind had already been doing ever since first becoming human by mastering fire-use and tool-making. Our pre*sapiens* ancestors took over for human use organic *materials* that would otherwise have been consumed by insects, carnivores, or bacteria. Ten thousand years ago, our horticulturalist ancestors began taking over *land* to grow crops for human consumption, displacing indigenous trees, shrubs, wild grasses, and various animals dependent on such plant life. At first the creatures displaced were those with teeth and claws instead of tools – nonhumans instead of other humans.

None of this is meant to justify European displacement of American Indians, Australian Aborigines, Polynesians, or Africans. But if we note what a common process takeover has been in the world's ecological history, we should be able to get beyond mere moral anguish and see clearly how fundamentally different that method was from another method by which human carrying capacity has most recently been stretched. The difference between the two methods has to be recognized if we are to understand our present predicament.

About 1800, a new phase in the ecological history of humanity began: Takeover continued, but now there was also *draw-down*. Machinery powered by combustion of coal (and, later, oil) enabled mankind to make and do things on a scale never before possible. Goods could be transported in larger quantity across greater distances. Tapping this "new" energy source eventually resulted in massive application of chemical fertilizers to agricultural lands; yields per hectare increased. Total land area applied to growing food for humans was substantially increased in the twentieth century by the elimination of draft animals and their requirements for pasture land. This time, the human carrying capacity of the planet was *supplemented,* not just enlarged, by digging up from nature's underground storehouse energy that had been captured from sunlight by photosynthesis in Earth's green plants long before this planet supported any mammals, let alone humans. Plants of the Carboniferous period grew, died, and became buried without the efforts of any farmers. The fact that no farm laborers or owners had to be paid to raise that vegetation and no investments in farm equipment had to be amortized, helped get us into our present predicament: The cheap remains of these prehistoric "crops" let us develop a social order based on energy prices that seriously understated true fuel values (costs of obtaining renewable fuel resources from crops grown today by human effort).

Carrying capacity was being augmented this time by drawing down a finite reservoir of prehistoric organic materials. The greatly expanded human opportunities allowed by draw-down would therefore be temporary. Since it was impermanent, this rise in apparent carrying capacity begged an important question. What happens if the ecological load increases until it nearly fills this temporarily augmented set of niches, and then, because the expansion is only temporary, the world finds itself (like the American Indians or the Australian Aborigines compressed within reservations) with an unsupportable population overload? The implications of a carrying capacity deficit for mankind's future are frightful (Cawte, 1978; Laughlin and Brady, 1978; Turnbull, 1972; 1978). What happens, for example, when oil deposits are so depleted that tractor fuel becomes unavailable or prohibitively expensive, and farmers again must divert one-fourth to one-third of the land on which they now raise food for humans to raising feed for draft animals?

Such questions are not asked when the world is viewed with a pre-ecological imperative and as long as the myth of limitlessness dominates political and economic thought. We tend not to notice that our fuel-burning mechanical slaves are becoming our antagonists. We and they now compete for limited space (Ognibene, 1980), and we and they must increasingly assert rival claims upon the land's limited capacity to produce food for our bodies or biomass fuel for our engines (Rask, 1981). We and they breathe the same air, and our engines can deplete the Earth's oxygen and add to its content of carbon dioxide and noxious or toxic combustion products faster than can our bodies.

Industrialization came about fast enough so that it enlarged per capita wealth and was not entirely devoted to increasing population. When modest enlargement of carrying capacity occurs over many generations, it tends to be used mainly to increase numbers. But if carrying capacity increments or supplements are enormous and come so suddenly that increase in population does not keep pace, they raise living standards. European takeover of the New World enlarged carrying capacity available to Europeans just fast enough to begin to have that salutary effect. By drawing down stores of exhaustible resources at an ever-quickening pace, industrialization (temporarily) augmented carrying capacity even faster. A marked rise in prosperity and a phenomenal acceleration of population growth were thus made possible. The dangerous myth of limitlessness was reinforced, and the earth's human passenger load increased by 400 million in only 65 years (Catton, 1980: 18–19) – a rate of increase of 14.28 percent per generation.

Unfortunately, this breakthrough we call industrialism differed in a fundamental way from earlier breakthroughs. It was based on a shift from takeover to draw-down; its increase of opportunities resulted from using substances that are not replenished in an annual cycle of organic growth. To believe in limitlessness in the 1980s is to believe that we can continue to find deposits of exhaustible resources as fast as we can use up what we've already found. Only once could the technologically most advanced peoples discover a second hemisphere to relieve pressure in their Old World; but modern industrial life is predicated upon massive and perpetual "exploration." It wrongly considers discovery of pre-existing but previously unfound deposits of mineral materials and fossil fuels "replenishment," calling these discoveries "additions to reserves."

The United States uses almost 10 kilograms of copper per year per person. Given the average level of metallic content of ores now being mined, Americans must *find* ore deposits each year containing some 227 million metric tons of ore (Lovering, 1969: 124) to sustain that rate of use. Present mineral and fuel substances used by industrial societies were deposited in the Earth's crust by geological processes over immense spans of planetary history. Unlike the growth cycle of crops or even of timber, geological renewal times for usefully concentrated deposits may be thousands or even billions of times longer than a human lifespan. Mankind cannot replenish the ores and fuels now being so ravenously extracted. The fact has to be faced that after ten millenia of cultural progress by which *Homo sapiens* has turned into *Homo colossus*, we have reverted to living as hunters and gatherers lived, depending now on massive extraction of substances only nature, not we, can provide. We have blinded ourselves to the problem by calling extraction "production." We thus pursue the holy grail of economic abundance by living – more or less unwittingly – according to the advertiser's cliche: "Hurry, quantities are limited. Get yours now!"

OVERSHOOT AGGRAVATED BY DEATH CONTROL

Less than three generations had enjoyed the benefits of industrialization when the role of microorganisms in producing diseases were discovered. Antiseptic surgery began in 1865 and was followed by an era filled with related breakthroughs in medical technology, hygenic practices, vaccination, antibiotics, and other fields. All served further to emancipate mankind from the life-ending effects of invisible little creatures for which human tissue formerly served as sustenance. Like other prey species newly protected from predators (see, e.g., Clark, 1964; Leopold, 1943), we have multiplied without regard for carrying capacity limits. Because of these achievements in the control of death, more of the unprecedentedly rapid rise in apparent carrying capacity resulting from industrial draw-down of resource stocks was devoted to supporting population increase, and less to supporting living-standard enhancement.

To the first three or four generations that experienced death control, this was of course a magnificent blessing. Parents were spared bereavement during their child-rearing years. Fewer children became orphaned. Adults were less often widowed in life's prime. People of all ages were spared the suffering and debilitation formerly inflicted by infectious diseases.

But these benefits helped us overshoot permanent carrying capacity. Industrialism had given us a temporary increase in opportunities – already a very dangerous blessing. Death control aggravated the unseen precariousness of our situation by releasing the brakes; a further acceleration of population increase followed, not linked to *any* increment of carrying capacity. Thus, between 1865 and 1975, world population grew at an unprecedented average rate of 27.55 percent per generation, loading the world with four billion precariously situated human souls. Yet many of those four billion continue to suppose ecosystem limits are unreal – either from ignorance of the carrying capacity concept or from naive faith that nothing that has changed need preclude simple extrapolation of past history (so that carrying capacity can be expected always to expand as needed). Further advances in technology are counted on before they hatch, and are expected to enlarge human carrying capacity, not reduce it. Indeed, enlargement of carrying capacity was technology's traditional function, so it is not easy to recognize that in the industrial era technology has magnified human appetites for natural resources, thereby *reducing* the number of us that a given environment can support.

LIGHT FROM BEYOND THE PIPELINE

It is not easy to acknowledge the human woe our future must hold because the fateful deepening of human dependence on nonrenewable resources and the imprudent human increase induced by their (necessarily temporary) abundance. *Homo colossus* plunges onward into a future whose central feature has to be an engulfing deficit in carrying capacity, but the temptation persists to attribute human hardship to such "forces" as inflation – that somehow "devours" our prosperity. Only the rising *monetary* costs of pursuing ever less accessible supplies are seen as obstacles to "the good life." Inexorably escalating prices, rather than physical depletion, are blamed for impeding lavish resource consumption. Tycoons and tyrants take the rap for inequities in distribution, while economic growth efforts desired even by the disadvantaged worsen the overall plight.

So let us think about Alaska, now a major source of the oil that feeds *Homo colossus* today. To build the pipeline from the North Slope to Valdez, the oil industry invested almost a hundred times as large a fraction of the modern gross national product (GNP) as the fraction of the 1867 GNP spent by the U.S. government to acquire Alaska from Russia. The forty-ninth state's greatest value, however, may well be not its oil or other extractable material resources, but some ecological lessons available from research conducted in some of its extreme environments.

For years, the Office of Naval Research supported investigations of arctic ecology through the Arctic Research Laboratory at Barrow, on Alaska's northernmost tip. There, the cyclical population explosions and crashes or die-off of lemmings have been under systematic scrutiny (Pitelka, 1958). The simplicity of biotic communities in that region results in periodic population fluctuations that are more easily seen than the less regular cycles occurring in more complex ecosystems. At Barrow, the brown lemmings (*Lemmus trimucronatus*) eat all year (beneath the snow cover in winter) and reproduce several times in a year. Because of the harsh climate, the vegetable matter they eat has less than one-fifth of a year in which to grow. Like us, therefore, they live by drawing down supplies of a resource that cannot replenish itself as fast as they can use it up. At the peak of a population outbreak, lemmings become as much as 500 times as numerous as in the year following a crash and preceding the next irruption. They crash or die off in large numbers because they destroy the resource base (simple, low-growing arctic vegetation) on which they depend. When the lemmings almost disappear, the vegetation renews itself – something the minerals and fossil fuels we use cannot do. Then the lemmings explode anew, destroy the vegetation once more, and crash again.

Ecological inquiry into the arctic are valuable to mankind because natural ecosystems in that region are so simplified by nature's own forces that they stimulate our situation and provide a clear preview of the destiny toward which *Homo colossus* hurls himself by unwisely daring to take the whole world for his oyster. We rush toward a lemming-like fate without pausing to discern it (Calhoun, 1971) because, under the influence of the myth of limitlessness, we mistake accelerated draw-down (which shortens our future) for a solution to our difficulties. It may be worth recalling that humans are now about 500 times as numerous as they were just before the horticultural revolution; and like the Alaska north coast vegetation, all of which becomes food for hungry lemmings, less and less of the world now escapes the ravenous jaws of technology.

The technological powers *Homo colossus* derives from the draw-down method enable us to practice the takeover method more ambitiously than ever. An ostensibly benign proposal for reducing human dependence on nonrenewable fossil fuels or hazardous nuclear reactors envisions taking over for human use enlarged fractions of the sunshine upon which the life of this planet depends. Energy shortage problems seem unnecessary to those who regard the vast amounts of unexploited sunlight falling on every square meter of the world's surface as solar energy that is now "going to waste." Proponents of this view seldom consider the possibility that taking over some additional fraction of the solar flux that now drives the world's climate might produce ecosystem changes detrimental to human interests. All that "unused" sunlight is assumed to be "available."

Indeed, small scale domestic applications of solar energy technology may be preferable to most other means of new energy conversion now under development; but it is ironic that solar energy enthusiasts criticize advocates of coal and fission for disregarding ecological costs and calculating only monetary costs, yet they glibly regard solar energy as "free" just because no corporation could bill us for the incoming sunshine.

They overlook major ways in which we already depend upon solar energy. It supports the agriculture that enabled our species to irrupt from a few million inhabitants in pre-Neolithic times to 500 times as many today. And it is not only through photosynthesis that solar energy supports agriculture. It also supplies the enormous power to evaporate each day from the surfaces of land and sea some 260 cubic kilometers of water, lift it up into the atmosphere, and make it available to rain down upon the farms, forests, and hydroelectric watersheds of the world.

Further, if 99.9 percent of the solar energy that reaches Earth's surface is not being captured by plants and fixed in organic molecules, this does not mean that solar energy is a "vast untapped reserve" awaiting man's exploitation. For mankind to try using even an additional one-tenth of one percent would be to impose upon the energy system of the ecosphere an extra load comparable to the energy budget of the entire standing crop of organisms of all kinds! How colossal should we aspire to be?

LAW OF THE MINIMUM

We mislead ourselves by concentrating too much on the *abundance* of sunshine, or of some other useful commodity (e.g., hydrogen in seawater, the fuel for "practically unlimited" energy in the dreams of fusion enthusiasts). Instead, according to a principle of agricultural chemistry formulated in 1863 by German scientist Justus von Liebig, whatever necessary resource is *least* abundantly available (relative to per capita requirements) determines an environment's carrying capacity. There is no way to repeal this "law of the minimum." But there is a way to make the application of Liebig's law less restrictive. People living in an environment where carrying capacity is limited by a shortage of one essential resource commonly develop exchange relationships with residents of another area in which there is a surplus of that resource. The second area has this surplus to exchange because demand for it there is limited, as its population is checked by a different resource deficiency. For example, on several volcanic islands in the southeastern part of the Fijian group (Thompson, 1949), yams grow well but timber does not, whereas on several nearby limestone islands, timber flourishes but not yams. Inhabitants of the two types of islands obtain what they lack by trading with each other. Timber from the islands where it grows well can thus provide fishing canoes for inhabitants of both sets of islands to harvest fish from the sea surrounding them all, and yams from the nonforested islands can nourish both populations.

Such trade does not repeal Liebig's law; it enlarges the scope of application of this law of the minimum. The composite carrying capacity of two or more areas with different resource configurations can be greater than the sum of their separate carrying capacities. Progress in transportation technology, and advancements in the organization of commerce (sometimes resulting from conquest or political consolidation), have thus enlarged the world's human carrying capacity. But if local populations (or lifestyles) transcend limits that used to be set by local scarcities, then they become *dependent* on trade (and upon the transportation technology and the organizational arrangements that make it possible) precisely to the extent that the composite carrying capacity of the combined environments exceeds the sum of the separate carrying capacities of the several localities when unconnected. (For an example, see Goldsmith, 1971: 15–16).)

In a resource-depleted future, if we face intensified competition between all of us (and thus between all distinguishable groups — nations, races, classes, occupations, religions), how likely is it that we will maintain whatever exchange relationships now provide a partial answer to Liebig's law? Pressures of competition, it must be expected, will erode the mutual accommodation and trust upon which organized patterns of exchange depend (see Turnbull, 1972; Yinger, 1976; Laughlin and Brady, 1978). One reason we are already in deep trouble is that those pressures easily make us suspicious of outgroups, persuade us to enact "protective" tariffs, or cause us to otherwise impede trade. By responding in this way we increase the harsh impact of Liebig's law. As we arm to defend our access to a resource such as oil, not only do we flirt with mutual destruction in response to mutual dread of deprivation, we also aggravate our mutual and unproductive draw-down of that very resource by diverting much of it to military consumption.

Extreme ecological difficulties can cause one human group to declare another group superfluous; genocide has occurred in response to such difficulties and could do so again (Catton, 1980: 220–221). It may indeed be the dire destiny of *Homo colossus*. More than is possible among animals of other types, humans develop social and cultural motives for resenting competitors. Cultural development has given humans the ability to hate and despise, and those destructive emotions tend to be aroused when other people differ from us, or compete with us. In severe ecological straits, the very qualities that distinguish mankind from other creatures may be our ultimate undoing.

We hope otherwise. But to extricate ourselves from the predicament we brought on by mistaking draw-down for just another useful breakthrough, we really need to begin weaning ourselves from our heavy dependence on exhaustible resources. To do this we must stop defining them as resources. We must learn to regard underground coal seams and oil pockets not as sources of fuel but as the best place to leave the enormous quantities of carbon that were so slowly withdrawn by nature from an atmosphere that could not otherwise have supported animal and human life. Can so radical a version of "conservation" be phased in? As trust in our fellow men and women erodes, can even the conventional forms of conservation be implemented? The time may be near when it will take an optimist to believe the future is uncertain.

PHOTO: United Nations

Passengers in Tokyo Central Station

REFERENCES

Allan, W. 1967. *The African Husbandman.* Edinburgh: Oliver and Boyd.

Borgstrom, Georg. 1969. *Too Many: A Study of Earth's Biological Limitations.* New York: Macmillan.

Boughey, Arthur S. 1975. *Man and the Environment* (2nd ed.). New York: Macmillan.

Calhoun, John B. 1971. "The Lemmings' Periodic Journeys Are Not Unique," *Smithsonian* 1(January): 6–12.

Catton, William R., Jr. 1976. "Can Irrupting Man Remain Human?" *Bioscience* 26(April): 262–267.

– – –. 1980. *Overshoot: The Ecological Basis of Revolutionary Change.* Urbana: University of Illinois Press.

– – –, and Riley E. Dunlap. 1980. "A New Ecological Paradigm for Post-Exuberant Sociology," *American Behavioral Scientist* 24(October): 15–47.

Cawte, John. 1978. "Gross Stress in Small Islands: A Study in Macropsychiatry," in Charles D. Laughlin, Jr. and Ivan A. Brady (eds.) *Extinction and Survival in Human Populations.* New York: Columbia University Press, pp. 95–121.

Clark, L. R. 1964. "The Population Dynamics of *Cardiaspina Albitextura* (Psyllidae)," *Australian Journal of Zoology* 12(December): 362–380.

Coale, Ansley J. 1974. "The History of Human Population." *Scientific American* 231(September): 41–51.

Day, Lincoln H., and Alice Taylor Day. 1964. *Too Many Americans.* Boston: Houghton Mifflin.

Durand, John D. 1967. "The Modern Expansion of World Population," *Proceedings of the American Philosophical Society* 3(June): 136–145.

Goldsmith, Edward (ed). 1971. *Can Britain Survive?* London: Tom Stacy, Ltd.

Laughlin, Charles D., Jr., and Ivan A. Brady. 1978. "Introduction: Diaphasis and Change," in Charles D. Laughlin, Jr., and Ivan A. Brady (eds.) *Extinction and Survival in Human Populations.* New York: Columbia University Press: 1–48.

Leopold, Aldo. 1943. "Deer Irruptions." *Transactions of the Wisconsin Academy of Sciences, Arts and Letters* 35: 351–366.

Lovering, Thomas S. 1969. "Mineral Resources from the Land," in Committee on Resources and Man (eds.) *Resources and Man.* San Francisco: W. H. Freeman: 109–134.

Ognibene, Peter J. 1980. "Vanishing Farmlands: Selling Out the Soil." *Saturday Review* 7(May): 29–32.

Pettersson, Max. 1960. "Increase of Settlement Size and Population Since the Inception of Agriculture." *Nature* 186(June): 870–872.

Pitelka, Frank A. 1958. "Some Characteristics of Microtine Cycles in the Arctic," in Henry P. Hansen (ed.) *Arctic Biology*. Corvallis: Oregon State College: 73–88.

Rask, Norman. 1981. "Food or Fuel: Will We Have to Choose?" *Worldview* 24(February): 5–7.

Thompson, Laura. 1949. "The Relations of Men, Animals, and Plants in an Island Community (Fiji)." *American Anthropologist* 51(April): 253–267.

Turnbull, Colin M. 1972. *The Mountain People*. New York: Simon & Schuster.

– – –. 1978. "Rethinking the Ik," in Charles D. Laughlin, Jr., and Ivan A. Brady (eds.), *Extinction and Survival in Human Populations*. New York: Columbia University Press, pp. 49–75.

Webb, Walter Prescott. 1952. *The Great Frontier*. Boston: Houghton Mifflin.

Wurster, Charles F. 1968. "DDT Reduces Photosynthesis by Marine Phytoplankton." *Science* 159(March): 1474–1475.

– – – . 1969. "Chlorinated Hydrocarbon Insecticides and the World Ecosystem." *Biological Conservation* 1: 123–129.

Yinger, J. Milton. 1976. "Ethnicity in Complex Societies: Structural, Cultural, and Characterological Factors," in Lewis A. Coser, and Otto N. Larsen (eds.) *The Uses of Controversy in Sociology*. New York: The Free Press: 197–216.

Economics and Sustainability: In Defense of a Steady-State Economy

Herman E. Daly

Herman Daly: Alumni Professor of Economics, Louisiana State University, author of Steady-State Economics *and numerous scholarly articles in the fields of population, development and resources.*

INTRODUCTION

A state of zero-growth can be achieved in two ways: as the failure of a growth economy or as the success of a steady-state economy. A growth economy and a steady-state economy are as different as an airplane and a helicopter. An airplane is designed for forward motion – if it cannot keep moving it will crash. Likewise our growth economy cannot be still without crashing into unemployment. We are currently witnessing the failure of growth economies in the face of increasing natural, social, and moral resistances to further growth. We have not yet converted our growth-bound economic airplane into a more maneuverable steady-state helicopter. What I want to advocate and discuss is zero growth under a steady-state economy, not a failed growth economy.

What is a steady-state economy? What reasons are there for believing that it is necessary and desirable? How could we move from where we are to such a state? These are the questions to be considered here.

WHAT IS A STEADY-STATE ECONOMY?

Neither the basic idea nor the reality of a steady-state economy is at all new. John Stuart Mill discussed the idea in his chapter "On the Stationary State" in his *Principles of Political Economy*. Throughout most of its tenure on earth humanity has existed in near steady-state conditions. Only in the past two centuries has growth become the norm. The steady-state economy is therefore a conservative notion, deeply rooted in both theorectical reasoning and historical experience. It is not a harebrained illusion like the ever-expanding economy of the post-Keynesians or the neo-Marxists.

The steady-state economy is defined by four characteristics. The first two are true of all steady states; the second two specify a particular kind of steady state, namely one that is sufficient, frugal, and sustainable:

1. A constant population – that is, a constant stock of human bodies
2. A constant population of artifacts – that is, a constant stock of physical wealth. Artifacts are essentially extensions of human bodies ('exosomatic organs', in A.G. Lotka's terms) so that 2 is a logical extension of *1*
3. The level or scale at which the two populations are maintained that is sufficient for a good life and sustainable for a long future
4. The rate of throughput of matter and energy by which the two populations are maintained that is reduced to the lowest feasible amount – that is, birth rates equal to death rates at low levels so that life expectancy is high, and production rates equal to physical depreciation rates at low levels, so that durability or 'life-expectancy' of artifacts is high

Throughput is the metabolic flow of matter and energy through the digestive tract of an economy. It begins with the input of resources, low-entropy[1] matter, and energy extracted from the natural environment (that is, depletion), and ends with the return of waste, high-entropy matter and energy back to the environment (that is, pollution). The throughput is a cost – the inevitable cost of replacement and maintenance – which should be minimized for any given scale of human and artifact population. Current practice in the growth economy comes closer to maximizing throughput because of its intimate connection to Gross National Product.

Note that the definition is in *physical* terms. Specifically, a steady-state economy is *not* defined as zero growth in GNP, nor does the definition imply constant technology, nor constant distribution of wealth and income. What happens to GNP is not important, but what happens to tecnology and distribution are important, and will be discussed later.

PHOTO: EPA – Documerica John White

37th and Prairie, Chicago, Illinois

Since this definition is a bit abstract, let us consider a concrete analogy: a steady-state library (actually advocated by some librarians as a way of coping with the exponential increase in the population of books). There would be a constant stock of books, limited, say, by storage space or maintenance budget. For every new book acquired, an old one must be gotten rid of – sold or recycled. Therefore a good librarian would never add a new book unless it were qualitatively better than some other book whose place it would take. The *quality* of the library would improve although the *quantity* of books remained constant.

Judging quality is often difficult, and there are two schools of thought among steady-state librarians. The 'consumer sovereignty' school says, get rid of the books checked out least often. The 'expert opinion' school says that consumer sovereignty would result in multiple copies of *Jaws*, while Dante and Sheakespeare would go to the shredder, and consequently that informed judgment of scholars should be the criterion. I will not try to resolve that debate. My only point is that a steady-state library is not static – on the contrary, qualitative improvement is forced by quantitative limits, which also provide incentives to develop new technologies, such as microfilming.

The service rendered by a library depends on the quality, size and intensity of use of the total stock of books, not on the rate of throughput or rate at which new books replace old books. The rate of acquisition of new books should be limited to the rate of physical and intellectual depreciation of old books; and the lower that rate is, the better off we are – the fewer trees have to be sacrificed for paper, and the less old knowledge has to be unlearned or discarded. We tend to think of intellectual depreciation as progress, but actually it is a *cost* of progress – often worth paying, but a cost nonetheless. If the old knowledge had been more solidly based and carefully constructed, then intellectual depreciation would be less.

Consider the economy as a library of artifacts and the analogies are evident. A steady-state economy, far from being static, is a strategy for forcing qualitative improvement and sustainability.

THE NECESSITY AND DESIRABILITY OF A STEADY-STATE ECONOMY

If a steady-state economy is necessary, then we must have it whether it is desirable or not. If it is desirable, then we should choose it even though we are not obliged to. Two relatively independent sets of arguments for the steady-state economy can thus be made.

A. Necessity

The eventual necessity of a steady-state economy follows from two basic principles: finitude and entropy.

1. *Finitude* – Contrary to the implication of some economic growth models, the best evidence from geographers is that the diameter of the earth in fact does *not* grow at a rate equal to the rate of interest! The solar flux is also constant, as are the rates of turnover of the basic biogeochemical cycles that support life on earth. The complex ecosystem, of which the human economy is a subsystem, seems to be a quasi-steady state: the total stock of matter is roughly constant, as is the throughput flow of solar energy. Eventually the economic subsystem must conform to the design principles of the total ecosystem of which it is a part; and the design principles appear to be those of a steady state, at least on the human time scale of generations, though not on the astronomical time scale of millions of years.
2. *Entropy* – Not only is the world finite it is also subject to the law of entropy. Not only must we recognize that the amount of petroleum

in the ground is finite, but also that we can burn each gallon only once. The entropy law tells us that recycling energy is always a losing proposition, and that recycling materials can never be one hundred percent complete. Depletion and pollution are inevitable costs that should be minimized but can never be eliminated.

Low entropy is the ultimate resource which can only be used up and for which there is no substitute. All substitution among resources is of one form of low entropy for another within a total set that is both finite and diminishing. Basically we have two sources of low entropy, solar and terrestrial. These differ significantly in their patterns of scarcity.

Solar energy is practically unlimited in total stock or amount, but strictly limited in its flow rate of arrival to earth.

Terrestrial minerals are strictly limited in total stock, but usable at a flow rate largely of our own choosing.

If all fossil fuels in the earth's crust could be burned, it would provide the equivalent of only a few weeks of sunlight – and the sun is expected to last for another four billion years. It is clear, then, which source of energy we must count on for the long run. Yet, contrarily, over the past century technology and substitution have brought about a shift from overwhelming dependence on the abundant solar source to overwhelming dependence on the scarce terrestrial source of low entropy. The reason for this shift, of course, is that terrestrial sources can be used at a rate of our own choosing – that is, rapidly – in order to fuel rapid economic and demographic growth.

But we are beginning to see the first phases of a reverse substitution, back towards solar energy, which, although it is the more sustainable source, cannot be used to fuel the continuation of growth based on the minerals bonanza of the past century. People no longer speak of converting petroleum into food, but of converting food products into petroleum substitutes. The populations of *homo sapiens* and *mechnistra automobilica* now enter into direct competition for the limited flow of solar energy annually captured on croplands.

Finitude and entropy imply the end of growth *sometime*, but what evidence is there that the time is now? Why not a thousand years from now?

Contrary to what one often hears, the most serious problem and the most convincing evidence of limits is provided by *renewable* rather than nonrenewable resources. Depletion of nonrenewables is to be expected, after all, since there is no point in leaving them in the ground forever if they are accessible. Yet there are two ways in which the too-rapid depletion of nonrenewables contributes to the destruction of the sustained yield capacity of renewables. First, high rates of depletion result in high rates of pollution of air and water which directly threaten biological resources. Second, rapid use of nonrenewables has allowed us to reach and sustain temporarily a scale of population and per capita consumption that could not be sustained by renewable resources alone. As our nonrenewable base runs out the danger is that we will try to maintain the existing scale (or worse, the customary rate of growth) by overexploiting renewable resources, thereby reducing permanent carrying capacity, and consequently reducing the total quantity and quality of life ever to be lived on earth.

Economist Lester Brown has summarized the evidence that the global per capita productivity of our four basic renewable resource systems has peaked and is now in decline (Brown 1979).

Forest productivity as measured by cubic meters per capita per year peaked in 1967 at 0.67 cubic meters.

Fisheries productivity, as measured by kilograms of fish caught per year per capita, peaked in 1970 at 19.5 kilograms.

Grasslands productivity is indicated by figures on wool, mutton, and beef. Wool peaked in 1960 at 0.86 kilograms; mutton in 1972 at 1.92 kilograms; and beef in 1976 at 11.81 kilograms.

Croplands productivity as measured by kilograms of cereals per capita per year peaked in 1976 at 346 kilograms.

Several things must be remembered in interpreting these numbers. The numbers reflect not only the effect of limits but also in some cases the effects of substitution. For example, wood production per capita may be down because of substitution of plastics for wood, not necessarily because of our having reached the limits of forest productivity. In like manner, reduced wool output per capita may reflect substitution of artificial fibers as much as limits to grassland carrying capacity. But even if that is so, it is sobering to remember two things. First, the existing levels of productivity are themselves gross overestimates of long-run, sustainable yield because they have been attained only with the aid of enormous mineral subsidies to mechanization, irrigation, fertilizers, insecticides, and transport. It is difficult to believe that existing levels of output can be maintained as we deplete remaining petroleum and mineral reserves, and as world population continues to grow for some time yet as a result of the momentum built into the youthful age structure. Second, it is significant that many substitutes (plastics and artificial fibers) are themselves petroleum derivatives.

Annual world petroleum output per capita rose from 1.52 barrels in 1950 to an all time high of 5.29 barrels in 1977. In 1978 it fell to 5.23 barrels. The petroleum bonanza is rapidly ending. The promise of a nuclear energy bonanza has proved a pipe dream, and a very dangerous one.

I conclude that the theoretical limits stemming from finitude and entropy are actual and observable in the world today. Further evidence of the increasing pressure of growing demands on dwindling capacities is provided by the acceleration of world inflation, with simultaneous unemployment.

A corollary to this conclusion is that the development plans of third world countries, which aim implicitly at a United States-style high mass consumption economy, are also pipe dreams. In addition to speaking of "underdeveloped" countries we must learn to speak of "overdeveloped" countries, which Charles Birch has defined as those countries which have per capita resource use rates that could not be sustained indefinitely if all people in the world consumed at that rate. The problem is not to convert underdeveloped into overdeveloped economies, but to convert both into steady-state economies at population and wealth levels that are sufficient for a good life and sustainable for a long future.

B. Desirability

We will consider two concepts of desirability: economic and ethical.

1. *Economic Desirability* – In the literal sense of "economic" physical growth is economic only as long as the marginal benefits of growth exceed the marginal costs. Once declining marginal benefits fall below increasing marginal costs we have "uneconomic" growth. Since we do not measure the costs of growth, no one can be sure that current economic growth is not really uneconomic growth. Even if it could be shown that at the margin the benefits of growth still exceed the costs, we must expect that eventually this situation will be reversed – at least if we believe in the law of diminishing marginal utility and the law of increasing marginal costs. And if these laws are rejected, then all of economic theory must be rewritten. In fact, these laws are simply common sense; they merely state that people satisfy their most pressing wants first, and that they first exploit the most accessible resources.

   Beyond some point, physical growth begins to cost more than it is worth. At that point a steady-state economy becomes economically desirable, even though it is not yet physically necessary.

2. *Ethical Desirability* – If sustainability is desirable, then the steady-state economy is desirable. Of course, nothing is sustainable forever; but there is a significant difference between twenty years and two-hundred-million years.

   Whence does the value of sustainability derive? If one is a pagan it might derive from a mystical respect for Mother Earth. If one is a Christian or a Jew, it derives from the duty of stewardship for God's creation which is declared intrinsically good independently of its critical instrumental value to its highest product, man. Most people, whatever their religious or philosophic background, seem to believe that it is wrong to treat the earth as a business in liquidation. There is at least a vague feeling that it ought to be handed on as a going concern, more or less as it was received, insofar as its capacity to support life is concerned. This can only be viewed as a religious insight.

Nevertheless, this perspective is not universal. Some people profess a naturalistic view and regard eco-castrophe as nothing more than ecological succession viewed with anthropocentric alarm. It is a matter of indifference to them whether or not human society is sustainable, since mankind, including his unknown successor, is no better than any other species. The consistent naturalist has no reason for being an environmentalist, other than mere personal tastes. At least the environmentalist must regard man as charged with a responsibility not applicable to other species: no one blames the deer for overgrazing. The naturalists who care not a whit for sustainability are logically consistent, but not numerous. The environmentalists who try to argue for sustainability on naturalistic grounds alone are numerous, but logically muddled. To be logically consistent the environmentalist must have a basically religious commitment to sustainability. There is no other answer to the logical naturalist who says that the sustainability of human society is nothing but an anthropocentric conceit.

One may believe that sustainability is desirable, yet be convinced that mankind will blow up the world within thirty years, and therefore consider it unwarranted to make any current sacrifice for a nonexistent beneficiary. In abandoning shallow chamber-of-commerce optimism one must be careful not to abandon hope as well and fall into the sin of despair.

If we agree that a steady-state economy is necessary and/or desirable, then we must ask, How can it be attained? Can we get there from here?

HOW TO ATTAIN A STEADY-STATE ECONOMY?

We may, of course, attain zero growth as a result of blind nemesis: Malthusain positive checks, overshoot and collapse, eco-catastrophe. The idea is to avoid that – to make a soft landing in the steady state rather than a crash landing.

I suggest two general principles as guidelines in thinking about institutions for a steady-state economy.

1. We must start from where we are with historically given initial conditions, and not assume an unrealistic clean slate. We have neither the time, nor the leadership, nor the wisdom to wipe out existing institutions and start over again with something radically different.

Present institutions of private property and the price system must be bent and reshaped, but not abolished. In some ways the reshaping will represent restrictions on these institutions, in other ways extenstions.

2. We should strive to attain the necessary macro-level stability and control with the least possible sacrifice of individual freedom and variability at the micro level. The macro is the level of control and predictability, the micro is the level of freedom, spontaneity and innovation; it is also the level at which most information exists.

I submit that there are three institutional changes which would allow us to move gradually toward a steady state while respecting the two principles. Briefly, they are: limits to population, limits to throughput, and limits to inequality.

A. Limits to Population

The right to reproduce must no longer be treated as a "free good" – it should be regarded as a scarce asset, a legal right limited in total amount at a level corresponding to replacement fertility, or less, distributed in divisible units to individuals on the basis of strict equality, and subject to reallocation by voluntary exchange. Such a scheme of transferable birth licenses or quotas has been discussed by a few economists and demographers (see Daly, 1977, pp. 56-61), but has not been taken seriously by many. Yet it combines macro control with micro freedom to a very high degree, and is a straightforward extension of the basic institutions of property and markets to cover a newly recognized scarce good.

Since many readers will not like this idea at all, perhaps a few caveats are in order. First, one is invited to substitute his own favorite plan, since the other two institutions do not depend on this particular population control scheme. Second, I admit that it might be wise to hold off on serious social control of population as long as the natural tendency is in the right direction. As the saying goes, "If it is working don't try to fix it." But I would keep this institution ready in reserve in case we are surprised by another baby boom, or in the more likely event that diminished carrying capacity leads us to recognize the desirability of a period of negative population growth.

B. Limits to Throughput

A constant population of physical wealth or artifacts is also a part of the definition of a steady-state economy. This population of things might best be limited indirectly by in essence limiting its "food supply" – limiting the throughput at the inflow or depletion end. The aggregate throughput of basic resources would be limited centrally, but the allocation of that limited aggregate among alternative uses would be left to decentralized decision-making, coordinated by the market. This might be done by the government setting aggregate depletion quotas on each basic resource and auctioning off the quota rights to individuals and firms; or alternatively, by a national ad valorem severance tax. Since energy is a kind of generalized resource necessary for the extraction of all other resources, we might begin by limiting energy throughput only.

This institution would bend the market in a restrictive manner: the determination of aggregate throughput would no longer be left to market forces. We cannot allow the market to set its own boundaries with the ecosystem, because the market can count only those costs and benefits that register themselves in terms of conscious, short run pleasure and pain, derived from exchangeable commodites. Most ecological interferences have costs or benefits that are organic in nature and longterm in effect, not subject to exchange and not associated with short-run pleasure or pain.

Nor is the market at all sensitive to the qualitative difference between poverty and luxury, which brings us to the third institution.

C. Limits to Inequality

Since the first two institutions rely significantly on the market it is necessary to correct the major defect of the market: the extreme inequalities in income that it tends to generate. I suggest something very simple: a minimum income coupled with a maximum income and a maximum on wealth – a limited band of inequality necessary for incentives, for rewarding work of varying irksomeness and intensity, yet ruling out extreme inequality. The minimum income could be financed from the revenue of the depletion quota auction or resource severance tax.

Since in the steady-state growth can no longer be appealed to as the answer to poverty (it has not been a very effective answer) it becomes absolutely necessary to achieve more fairness in the distribution of income. Limiting inequality to an acceptable range represents a restriction on the market's role as distributor of income. This also implies imposing some limits on size and power of corporations.

There is much more to be said about each of these institutions, but the general outline should be evident. It is important to propose something specific, even if incomplete, in order to oblige critics to come up with something better, which no doubt they will do, once they learn to take the matter seriously.

I am afraid that, to date, the reaction of many economists has been not to take the issue seriously. Their reaction is reminiscent of the attitude taken by an early American economist named Daniel Raymond to the ideas of Malthus. In his 1820 treatise, *Thoughts on Political Economy,* Raymond explains to his readers why he has omitted any serious consideration of Malthus' ideas:

Although his [Malthus'] theory is founded upon the principles of nature, and although it is impossible to discover any flaws in his reasoning, yet the mind instinctively revolts at the conclusions to which he conducts it, and we are disposed to reject the theory, even though we could give no good reason.

I believe that this attitude prevails today in the reaction of many economists to limits to growth arguments, but without the saving grace of Mr. Raymond's naive honesty.

[1]The Entropy Law (second law of thermodynamics) can be explained in relation to the law of conservation of matter-energy (first law of thermodynamics) by analogy to an hourglass. The sand in the glass represents matter-energy. The hour glass is a closed system, no sand enters or leaves (first law). But there is a qualitative difference between sand in the top chamber and that in the bottom chamber. The upper chamber sand has the potential to fall (do work) and that in the lower chamber does not. The second law (Entropy Law) says that sand in the upper chamber always diminishes when work is done, while that in the lower chamber is always accumulating, and that the hourglass cannot be turned upside down. The physicist Erwin Schroedinger said that organisms live by sucking low entropy from their environment – that is, by importing low entropy matter-energy and exporting high entropy matter-energy. The same is true of economies.

REFERENCES

Brown, Lester R. "Resource Trends and Population Policy: A Time for Reassessment," *Worldwatch Paper No. 29,* May, 1979, Washington, D.C.

Daly, H.E. 1977. *Steady-State Economics.* San Francisco: W.H. Freeman and Co., Publishers.

Daly, H.E., ed. 1980. *Economics, Ecology, Ethics.* San Francisco: W.H. Freeman and Co., Publishers.

SECTION TWO

HEARTLAND

Sacred Land, Sacred Sex

Dolores LaChapelle

Having lived most of her life in the mountains of Utah and Colorado, Dolores LaChapelle combines life and work by specializing in various forms of experiential education within the context of the natural environment. She designed and directed an elementary educational program through outdoor experience for the public school system in Aspen, Colorado. She also designed an ecological awareness course for Nova and founded and taught in a Community School in Alta, Utah. Her publications include two books: Earth Festivals *(1976) and* Earth Wisdom *(1978).*

INTRODUCTION

This essay explores the relationship between human sexuality and the natural world around us. Others, such as Wilhelm Reich, Norman Brown and Herbert Marcuse, have begun with the individual and then gone on to question the sexual mores of our culture. This approach is too anthropocentrically narrow for my purpose. Instead I go to the roots of the matter by considering three different areas—each of which recognizes the reciprocal relationship of human and non-human: first, primitive cultures; second, Taoism, with its direct ties to the primitive; and third, D.H. Lawrence, the only modern literary figure who has ventured into this terrain of human sexuality and nature.

This complex study may leave the reader with more questions than answers but the first step in our search for a modern ecological consciousness is learning to ask the right questions. We have scarcely begun to do that.

> *Oh, what a catastrophe, what a maiming of love when it was made a personal, merely personal feeling, taken away from the rising and the setting of the sun, and cut off from the magic connection of the solstice and equinox! This is what is the matter with us, we are bleeding at the roots, because we are cut off from the earth and sun and stars, and love is a grinning mockery, because, poor blossom, we plucked it from its stem on the tree of Life, and expected it to keep on blooming in our civilized vase on the table.*
>
> *D.H. Lawrence*

Today, more than half a century later, it is even more obvious that we are "bleeding at the roots"; yet, judging from the dozen books published last year on the fiftieth anniversary of Lawrence's death, the full implications of this "love," which he spent his lifetime exploring, are as little understood as they were in his own day. Of his contemporaries only the perceptive Scandinavian novelist, Sigrid Undset grasped the importance of Lawrence's work. In 1938 she wrote: "So many-sided was Lawrence and so intensely did he live his life that he became a representative figure – the man of mystery who symbolized his civilization at the moment when it reached a crisis. It is among other things a crisis of population and an economic crisis. In the language of mythology it means that the phallus has lost its old significance as a religious symbol... Much of what is happening in Europe today and yet more that will doubtless happen in the future are the brutal reactions of mass humanity to the problems which the exceptional man, the genius, D.H. Lawrence, perceived and faced and fought in his own way."

The following year, 1939, the "brutal reactions" which Undset so clearly anticipated began with Hitler's invasion of Poland, quickly escalating until the slaughter became world wide; yet, once the war was over it was "back to normal" with an unprecedented "baby boom" as fears and uncertainties about the future were covered over with an emphasis on settling down and raising a family. It was "the world all in couples, each couple in its own little house, watching its own little interests, and stewing in its own little privacy" as Lawrence so graphically stated in *Women in Love.*

In the 1960s when the products of this middle class "baby boom" reached maturity, they rebelled against the affluent "system" bequeathed to them, confusedly proclaiming themselves pro-life and against exploitation of that life in any way. In trying to help the victims of exploitation such as the blacks and the poor they discovered that they, too, were powerless against the "system"; while in experimenting with freer, uncommitted forms of sexuality, they frequently found such sex meaningless. Furthermore while working alongside their men in these crusades, the women began to realize that they were as exploited as any other minority. Out of this impasse came the many aspects of the "woman's movement" with its confusion over the "rights" of man *versus* woman. We are not nearer a solution than in Lawrence's time. There is more hostility, more anger and more hatred between the sexes. We are obviously asking the wrong questions if we are getting such devastating "solutions."

When Undset mythologically defined the crisis which our culture is facing in terms of the phallus losing its old significance as a religious symbol she was referring to our European cultural inheritance of two thousand years of Judeo-Christian emphasis on sex for procreation. To get to the roots of what sex is for among human beings we have to go beyond this temporary anomaly of two thousand years of Christianity to the total span of the 300,000 years of *Homo sapiens'* life on earth. When we do this we discover that, as in so many other aspects of human behavior, we got our basic sexual patterns from our ancestors, the higher primates.

The higher primates made the break-through from the usual mammalian sexual pattern in which all the females come into heat at the same time of the year thereby creating intense rivalry among the males during this limited time. Usually the dominant male secures a harem of females leaving the other males to wander alone or rove in "bachelor" bands. Such activity effectively breaks off any continuity of relationship among the entire herd or tribe. In the higher primates all this is changed. With females coming into heat throughout the year, at any one time some females are always available so copulation becomes an on-going activity. In fact Schaller says that gorillas show no sexual jealousy whatever. Mating becomes a year round possibility, and sexual activity becomes a method of creating closer bonding rather than a temporary breaking-up of society as in most mammalian species. This continuous sexual atmosphere leads to social relationships based on sexual status, and ritualized forms of sex becomes a form of communication about other matters. Among higher primates sexual postures such as mounting and presenting are examples of non-reproductive sexuality. Many different kinship ties develop as well as various relationships to consorts of the opposite sex. Sexuality becomes a general style and way of behaving throughout the primate society. Freeing sexual patterns from procreation alone increases the bonding within the group.

Sexual activity continued to serve as the major bonding mechanism for human tribes of hunter-gatherers. For over 99% of the time that human beings have been on this earth they have lived as hunter-gatherers. Only in the last 10,000 years have they begun to live off agriculture. Ten thousand years is only four hundred human generations. In so few generations there is no possibility of real genetic change. "We and our ancestors are the same people," as the anthropologist, Carleton Coon, succinctly pointed out; therefore the same physiological and psychological structures form the roots of our behavior pattern today.

Until quite recently agriculture was considered to be an enormous step forward for the human race, one which vastly improved mankind's life; but during the last fifteen years new areas of research have made this idea seem very dubious indeed. Originating with the 1967 conference on "Man the Hunter," the proliferating research since then has clearly shown the advantage of the hunting-gathering life over both the agricultural life and modern industrial culture. Even today modern hunting-gathering cultures such as in the Kalahari desert, work an average of two days a week to secure all their food leaving vast amounts of leisure for the preferred human pursuits of dancing, music, conversation and art. Recent research has proved that hunting and gathering provided higher quality and more palatable food than agriculture; furthermore, crop failures cannot wipe out the entire food supply of hunter-gatherers because their food supply is so diverse. Only ten thousand years ago a few groups began agricultural practices, yet by two thousand years ago the overwhelming majority of human beings lived by farming. What happened to cause this incredible shift to agriculture all over the earth in only 8000 years? Although clues have been accumulating during the past few decades not until 1977 did the answer become clear when Mark Cohen wrote *The Food Crisis in Prehistory* (Cohen 1977).

Drawing on more than eight hundred research studies Cohen shows that human populations grew so large that hunters caused the extinction of great numbers of large mammal species by the end of the Pleistocene, thus forcing large numbers of human beings to resort to agriculture. Cohen points out that the only advantage that agriculture has over the hunting-gathering life is that it provides more calories per unit of land per unit of time and thus supports denser populations. He explains that fifty species of large mammals were extinct by the mid-Pleistocene in Africa, two hundred species in North and South America by the end of the Pleistocene, and in Europe the enormous herds of grazing animals were gone by about the same time. Paleolithic hunters had destroyed the easiest, most tasty protein source. Mythologically, we can say that the hunters realized that the Mother of all Beasts no longer sent her animals among them for food.

To deal with this new situation, over a period of several thousand years, human beings developed several strategies – some of which led to "biosphere" cultures and others to "ecosystem" cultures. One of these strategies, agriculture, required more work than hunting-gathering, thus encouraging larger families of children to provide more workers, which in turn, meant more intensive agriculture and so on in an ever increasing spiral of scarcity, hard work and destruction of soils. Eventually, this led to enslaving other peoples as workers. These conquerors, the "biosphere" people, as Gary Snyder so succinctly explains, "spread their economic system out far enough that they [could] afford to wreck one ecosystem, and keep moving on. Well, that's Rome, that's Babylon" and every imperialistic culture since then.

Biosphere cultures assumed that Nature was no longer the overflowing, abundant Mother, giving all that humans needed. She had withdrawn her plenty; the never-ending stream of animals was gone. Nature was not to be trusted anymore; therefore humans must take affairs into their own hands. Within the last 500 years of the 10,000 year agricultural age, all the world's so-called "great ethical systems" arose, beginning with Confucius and Buddha (approximately 500 B.C.), through the Hebrew prophets and Plato, ending with Christianity, in the beginning of our present era (the year 1 A.D.). What we really have here is the establishment of religious systems based on "ideas" out of the head of individual human beings – Buddha, Moses, Jesus, St. Paul and others.

Before exploring the second type of culture – the "ecosystem" cultures – it is necessary to take a look at the two basic, underlying ways of looking at life. One emphasizes the initial point of the process of life – the birth or coming into being of new things and the other emphasizes the end of the process, the death or disappearance of old things. The major world religions mentioned above, the Judeo-Christian tradition and Buddhism, focus on the final phase, the inevitable decline, destruction and death of all things. Buddhism teaches the way of deliverance from the wheel of rebirth – from the impermanence and ongoing passing away of all things. This universal impermanence of things arouses an ultimate anxiety in the rational mind and, according to Toshihiko Izutsu, it is this "existential pessimism that is at the bottom of Buddhism." Christianity, as well, focused on the endings of things – the transiency of life, preferring to concentrate on the "ideal" of life after death in a perfect state called heaven. In these "ethical" world religions, life on this earth or on this "plane of being" is transitory, unimportant, even illusory; therefore the earth itself becomes expendable, of no real value thus permitting total exploitation of nature.

The other world view is exemplified by Taoism which focuses on the initial point of the process, the birth, the coming into being of new things – "the never-ceasing procreativeness" of life. The word, *I* in the Chinese classic, the *I Ching* means just this – "the eternal continuation of new interconnected lives," according to Izutsu. There is no separation between human and non-human. The human mind came out of nature and is identical with the structure of the cosmos. Nowhere in Taoism is there a "great man" who formulated "ideas" out of his head as in ethical religions. The legendary Lao Tzu and the historical figure, Chuang Tzu are the most noted exponents of the Tao; but it is clear from their writings that nature was considered much too complex for the human mind to impose order on it. Human society could only be brought into order by fitting itself into the order of Nature because human society was only a smaller part of the whole of the Tao.

With the distinction between these two viewpoints on life clarified we can now turn to "ecosystem" cultures. We find that instead of taking up agriculture these people moved off into marginal areas – high mountains, deserts, deep jungles or isolated islands and learned to pay attention – to watch carefully and to reverence all of life for it was their body, their life. They developed rituals which acknowledged the sacredness of their land; thus enabling them to remain aware of the sacred cycles of taking life to live but also of giving life back so that the whole of the land could flourish not just the small segment of that whole – the human beings. Because their economic basis of support consisted of a limited natural region such as a watershed, within which they made their whole living, it took just a little careful attention to notice when a particular species of animal or plant became scarcer and harder to find. At such times they set up taboos limiting the kill. They began to understand that they could destroy all life in their environment by excess demands on it if there were too many human beings there; thus they came to understand that sex, too, was part of the sacred cycle. Misused it caused destruction not only within the human tribal group but on all the life around them. Used with due reverence for its power it brought increased energy and unity with all other forms of life.

In the primitive cultures which developed out of the "ecosystem" way of life, based on "sacred land, sacred sex," much of the wisdom of the tribe was devoted to "walking in balance with the earth." Human population was never allowed to upset this equilibrium. Early ethnologists in North America found that the child spacing and child rearing practices growing out of this approach to the earth showed remarkable similarities in tribes as far apart as the Arctic and the Southwestern desert. For instance, John Murdock wrote: "Infanticide is frequently practiced among the Eskimo of Smith Sound, without regard of sex...The affection of parents for their children is extreme...the children show hardly a trace of the fretfulness and petulance so common among civilized children, and though indulged to an extreme extent are remarkably obedient. Corporal punishment appears to be absolutely unknown." The Pima Indians in the southwest, according to Frank Russell, regulated births in this manner: "Babies were nursed until the next child was born. Sometimes a mother nursed a child until it was six or seven years old and if she became pregnant in the meantime she induced abortion by pressure upon the abdomen...the youngsters are seldom whipped." In the *Assiniboin* tribe, according to Edwin Denig, the children were never hit or corrected in any way. "Notwithstanding this they are not nearly as vicious as white children, cry but little, quarrel less, and seldom if ever fight." He continued by saying that infanticide is very common among the Assiniboin, the Sioux and Crows. Cushman reported that, "Children were never whipped among the Choctaw, Chickasaw and Natzhez Indians." Infanticide or abortion together with the lack of punishment of children and the quiet, "good" children proved a continual source of amazement to these early research people. Among these tribal people however, the birth of a child was not an "accident" left up to the individual parents, but instead, was regulated by ritual or contraception or abortion so that the particular child would fit into the overall health and stability of the entire ecosystem – the tribe plus the rest of the community consisting of the soil, the animals and the plants. This is shown in a chant from the "Reception of the Child" ritual among the Omaha Indians:

> *Ho! ye Hills, Valleys, Rivers, Lakes, Trees, Grasses,*
> *all ye of the earth, I bid you hear me!*
> *Into your midst has come a new life! Consent ye,*
> *I implore!*
>
> * * *
>
> *Ho! Ye Birds great and small that fly in the air,*
> *Ho! Ye Animals great and small that dwell in the*
> *forest,*
> *Ho! Ye Insects that creep among the grasses and burrow*
> *in the ground – I bid you hear me!*
> *Into your midst has come a new life! Consent ye, I*
> *implore!*

In some cultures the food/reproduction/energy cycle is so attuned that there is no need for conscious birth control. The !Kung have lived in the Kalahari desert for at least 11,000 years. Recently because of the drop in the water table they have been forced to begin moving near the farming villages of the settled Bantu. Here the !Kung women are losing their equal status, children are becoming more aggressive and the once stable population is exploding. Population stability among the nomad !Kung is mostly due to the scarcity of fat in their diet which delays the onset of menstruation. First babies are usually born when the mother is about nineteen. They are nursed for four years and the mother rarely conceives during that time because nursing women need one thousand extra calories per day; therefore she does not have enough fat left over for ovulation to take place.

For these "ecosystem" people it is not possible to speak of the human beings' relation to the universe but rather of a universal interrelatedness.

Man is not the focus from which the relations flow. For instance, Dorothy Lee, in her study of the Tikopia natives found that "an act of fondling or an embrace was not phrased as a 'demonstration' or an 'expression' of affection — that is, starting from the ego and defined in terms of the emotions of the ego, but instead as an act of sharing within a larger context." By way of illustration there follow three detailed examples: in the Ute Bear Dance sex was used to bond the widely scattered hunting bands into the tribe as a whole; in the Eskimo game of "doused lights" sex was used as an emotional cathartic and in the final example, a modern Odawa Indian shows that the sharing of sex still contributes to the bonding of the tribe.

During most of the year the Ute tribe was split up into small kinship groups hunting in widely separated parts of the high Rocky Mountains. Once a year the entire tribe met for the annual spring Bear Dance. They waited for the first thunder, which they felt awakened the hibernating bear in its winter den and awakened the spirit of the bear in the people. A great cave of branches, the *avinkwep* was built with the opening facing the afternoon sun. At one end of the cave a round hole was dug to make an entrance into a small underground cave. Over this area the resounding basket was placed with the notched stick resting on top. When played this made a sound like thunder "spreading out over the awakening land and rumbling in the spring air." The singers closed in around this thunder and the dance began. Because the female bear chooses her mate, the woman chose which man she would dance with by plucking his sleeve. For three days the dance continued. The spirit of the bear filled the *avinkwep*. From time to time a couple would leave the dance and "take their blanket up into the brush of the hillside to let out the spirit of the bear and the thunder of spring that had grown too strong in them." Many healings took place during the Bear Dance. At noon of the third day the Dance ended and gradually over a period of days the big camp broke up as the small hunting groups went out into the hills. A woman who plucked the sleeve of a man during the Bear Dance might visit the bushes with him for an hour, or for the entire night or might stay with him for the entire year's hunting until the next Bear Dance, or even for "many snows." Here ritualized, sacred sex served the function of putting the individuals together again as a tribe and back in connection with their land through their totem animal, the Bear.

Peter Freuchen tells of an Eskimo game, "doused lights" where many people gathered together in an igloo. Lights were all extinguished and there was total darkness. No one was allowed to say anything and all changed places continuously. At a certain signal each man grabbed the nearest woman. After a while, the lights were lit again and now innumerable jokes could be made over the theme: "I knew all the time who you were because..." This game served a very practical purpose if bad weather kept the tribe confined for a long time because the bleakness and loneliness of the Arctic become difficult to face at such a time. The possibility of serious emotional trouble is ever-present because bad weather can mean little food or an uncertain fate but after this ritualized sexual game is over when the lamp is lit again the whole group is joking and in high spirits. "A psychological explosion – with possible bloodshed – has been averted," Freuchen explains.

Wilfred Pelletier is a modern Odawa Indian who left his island reserve in Canada and became a success in the white man's world but he found it wanting and returned to his reserve. He says that his own introduction to sex was provided by a relative. "I still look on that as one of the greatest and happiest experiences of life. From that time on, it seems to me that I screwed all the time, without letup. Not just my relatives, who were not always available, but anywhere I could find it, and it always seemed to be there... On the reservation people were honest about their feelings and their needs, and as all the resources of that community were available to those who needed them, sex was not excluded. Sex was a recognized need, so nobody went without it. It was as simple as that."

While in the previous examples sexual activity served as a bonding mechanism for the tribe, in the Tukano tribe their entire ecologically sophisticated cosmology is largely derived from the model of sexual physiology. The Tukano live in the rain forest of the northwest Amazon Basin with its difficult climate and easily depleted natural resources. In their world-view the Sun-Father, a masculine power fertilizes the feminine element, the earth. Creation is continuous because the energy of the sun, which to the Tukano is seminal light and heat, causes the plants to grow and mankind and animals to reproduce. This procreative energy of the sun flows in limited quantity continuously between all parts of the universe. Humans can remove only what is needed for their life and only under particular ritual conditions. Whatever energy is borrowed must be put back into the circuit as soon as possible.

They have little interest in new knowledge about how to more effectively exploit their environment; instead their basic concern is knowing more about the natural entities around them and finding out what the beings of their world need from them. They feel that they must fit themselves into the ongoing circuit of energy if it is to continue. They have detailed knowledge of their microclimate. Reichel-Dolmatoff of the University of the Andes in Columbia, who studied them, discovered that they understand such phenomenon as "parasitism, symbiosis, commensalism and other relationships between co-occurring species" and that their shamans have consciously adapted some of these strategies into human life. Their myths tell of particular animal species which became extinct or were punished for such behavior as gluttony, aggressiveness and other kinds of overindulgence which might upset or stop the energy flow through the entire circuit.

The birth rate is controlled by the use of oral contraceptives and sexual continence. Certain herbs are eaten which cause temporary sterility thus providing proper spacing between children so that the eldest has the advantage of a totally adequate diet. The number of children for each couple is quite low. Sexual abstinence is required before any ritual. Food and sex are closely related and symbolically equivalent.

The Sun-Father created a limited number of plants and animals and put them under the protection of the Master of Animals, considered to be a spirit-being with phallic attributes. In order to obtain the Master of Animal's permission to kill an animal, the hunter must undergo rigorous preparations including fasting, sexual continence, cleansing by bathing and emetics. Plant and earth resources such as thatch for a roof or clay for pots cannot be gathered until certain ritual actions are completed which bring with them the necessary permit from the spirit owners of each resource.

On ritual occasions the shaman does not merely ask for abundance of food; instead the ritual provides him with the occasion for a sort of stock taking of the available resources and, according to Reichel-Dolmatoff "can be viewed as rituals concerned with resource management and ecological balance." The Tukano do not see nature as something apart from man so there is no way that human beings can confront it or oppose it or even try to harmonize with it. The individual human being, however, can unbalance the cosmic system by his personal malfunctioning as a sub-part within the overall system; but he can never stand apart from it.

Generally, in the past a linear cause and effect model has been used to explain how cultures function. "Only in the last decade," Reichel-Dolmatoff points out, "ethnographers and archaeologists are coming to accept as the only kind of explanatory model which can be used to handle ecological relationships the kind of overall systems model which was adopted by 'primitive' Indians a very long time ago."

The linear, cause and effect model, with its dualistic presuppositions has been inherent in Western civilization since the time of the Greeks. Only since the famous Macy Conferences (1947-1953), which "invented cybernetics," has the West been moving toward the "systems model" of thinking. In the East, however, the "systematic" way of thinking dates from the Taoists of very ancient times. Laszlo, a pioneer of systems thinking in Western philosophy points out its advantages when he states, "In opposition to atomism and behaviorism, the systematic view of man links him again with the world he lives in, for he is seen as emerging in that world and reflecting its general character." Or, as the Chinese would say: "Heaven, Earth and Man have the same *Li*."

Painting by Ed Roxburgh

Taoism does not consider that the order in nature comes from rules laid down by a celestial lawgiver as in our Western "laws of nature" concept; but from the spontaneous cooperation of all the beings in the universe brought about by following the *li*, the pattern of their own natures. The earliest meaning of the word, *li*, came from the pattern in which fields were laid out for cultivation in order to follow the lay of the land. Hence the earth itself was the ordering principle for a particular place. Generally, in ancient times the word, *li*, was used to describe the pattern in things such as the markings in jade or the fibres in muscles. The word, *li*, eventually came to be used as the principle or organization in the universe. It is the order to which parts of the whole have to conform by virtue of their very existence as parts within that whole. If they do not behave in a particular way (according to their *li*) they lose their relational position in the whole and become something other than themselves. This is the underlying reason for the elaborate seasonal rituals practiced throughout Chinese history. Since human beings were considered an integral part of nature, to follow their own human *li* required that they conform to the ongoing seasonal pattern, the *li* of nature.

Early European translations of the Taoist classics distorted the meaning of key Taoist concepts by attempting to fit the Chinese words into Christian metaphysical terms, which resulted in an other-worldly, "spiritualistic" text. Recently, the work of a number of men such as the English scholar, Joseph Needham and the Japanese scholar, Toshihiko Izutsu has succeeded in freeing Taoism from its Christian overlay and tracing its roots directly back to the primitive shamanistic cultures existing on the frontiers of the ancient Chinese Empire. Izutsu observes that the mythopoetic imagery in Chuang Tzu's writings is of shamanistic origin and further states that Chuang Tzu was a "philosophizing shaman."

Because of the very early development of writing in China not only does Taoism provide us with writen material coming directly out of the original primitive concept of man within nature uncorrupted by any later "ethical-religious" systems but also gives us the advantage of centuries of further development and elaboration of these concepts within the most "civilized" culture in the world. According to Taoism the energy of the universe operates through the interaction of yin and yang (female and male); thus Taoism developed the most sophisticated methods of any culture for dealing with sexual energy within and between humans and nature. Until recently, Taoist sexual techniques were dismissed as superstition but these methods were rescued from obscurity when Joseph Needham devoted a considerable section of the second volume of his life-long work on *Science and Civilization in China* to the subject.

Needham explains that, "The purpose of the Taoist techniques was to increase the amount of life-giving *ching* as much as possible by sexual stimulus... Continence was considered not only impossible, but improper, as contrary to the great rhythm of Nature, since everything in Nature had male or female properties. Celibacy (advocated later by the Buddhist heretics) would produce only neuroses." Taoist sexual techniques were used to increase the energy not only between man and woman but within the human group as a whole and between the humans and their land. From these relationships came the exquisite sex/nature poetry of China.

The most important ritual was called "The True Art of Equalising the Chhi's" or "Uniting the Chhi's" of male and female, dating back to at least the second century A.D. Most of what we know of these ancient rituals comes from a convert from Taoism to Buddhism who wrote the Hsiao Tao Lun or "Taoism Ridiculed," in much the same fashion as pagan converts to Christianity wrote ridiculing pagan customs. The ritual of "Uniting the Chhi's" occurred on either the nights of new moon or full moon after fasting. It began with a ritual dance "coiling of the dragon and playing of the tiger" which ended either in a public group ritual intercourse or in a succession of unions involving all those present in chambers along the sides of the temple courtyard. A fragment of the highly poetic book of liturgy for this ritual called the Huang Shu survived. During the Ming dynasty most of these Taoist sexual ritual books were destroyed. Fortunately some were preserved becaused they had been translated by the Japanese in about the tenth century.

In both the Chinese sexual rituals and in primitive tribal rituals sex itself was made numinous. In the Chinese rituals it is not clear what deities were worshipped but Needham refers to Maspero who says that they seem to have been star-gods, the gods of the five elements and the spirits residing and controlling the various parts of the human body.

Buddhist asceticism and Confucian prudery were both scandalized so that there were no public Ho-Chhi festivals after the seventh century. Private practice continued well into the Sung dynasty in Taoist temples and, among certain classes of lay people, until the last century. While Needham was in China in 1945, gathering up Taoist documents in danger from the revolutionary struggles he visited many old Taoist centers dating back as far as 554 A.D. This enabled him to get considerable insight into the ideas behind Taoist sexual techniques. When he asked "one of the deepest students of Taoist" at Chhêngtu, "How many people followed these precepts?" The answer was: "Probably more than half the ladies and gentlemen in Szechuan."

To understand why the Taoists devoted so much attention to sexual techniques it is necessary to understand the importance of the structure of the pelvic region with its central bony mass, the sacrum. Because Western culture puts so much emphasis on the rational mind (specifically, the rational, left hemisphere of the neo-cortex), the Taoist insistence on the importance of the lower mind, located four fingers below the navel (*tantien* in Chinese and *hara* in Japanese), seemed utter nonsense until quite recently. In the nineteen twenties when D.H. Lawrence emphasized the "solar plexus" he was riduculed; but, in the last two decades with the growing popularity of such disciplines as Tai Chi, Aikido and other martial arts as well as therapies such as Rolfing, the pelvic area is coming to be recognized as truly our "sacred middle" – the area within us where the flow of energy takes place between us and the cosmos. The functions of sex, prenatal life, birth, assimilation of food as well as deep emotions all take place in this area. "Sacred Middle" refers to the sacrum, the bony plate which gets its name from the same Latin root as the word sacred. The lower five vertebrae, which in the adult become fused into a single, curved, shield-shaped plate, make up the sacrum. Since the muscles of the entire pelvic girdle are attached here, this is the area which makes us human. The enlarged human brain developed only after we achieved true upright posture. Furthermore, all the muscles involved in walking, standing and sitting converge here. Of these muscles, the psoas muscles are the most important ones for determining the human, upright position; while the pubo-coccygeus muscles, which attach the legs to the inner sides of the spine and run on up to fasten the pelvic rim to the front of the ribs and the breastbone, literally hold us together. Both these large strong sets of muscles crisscross in the pelvic area around the sexual organs.

Much of the corrective work accomplished through Rolfing has to do with breaking down the overly rigid abdominal muscles so prevalent in our culture. The psoas muscles are then freed to function thus permitting the body to realign itself with gravity. It has been proved through laboratory experiments that rolfing these muscles cuts down and often completely eliminates the state of underlying anxiety so prevalent in our culture. This connection between what we generally conceive of as a "mental" condition and the actual facts of its cure through manipulation of muscles within the pelvic region defies the usual Western logic; yet the Chinese have always considered the *tantien*, the pelvic region as the location of the "other" brain. Not only is the *tantien* the seat of strong emotions but, when trained by Tai Chi it can also sense these emotions in other people as well as registering currents within the earth. Modern experiments with dowsers prove that the pelvic area is the particular physical location within the dowser's body which indicates water beneath the earth.

The most ancient of the martial arts, Tai Chi was developed in the Taoist Shaolin monastery in the mountains of China. Each of its 108 forms deals specifically with the muscles discussed above. It is these same muscles, which Tai Chi liberates, that are crucial for the practice of Taoist sexual techniques. For the man, these muscles are essential for the Taoist technique of orgasm without ejaculation. For the woman, during the moment of vaginal orgasm, the fascial and coccygeal muscles of themselves, with no other effort, can prevent the entry of semen and thus provide an automatic method of birth control when needed. This latter technique may be the answer to some of the puzzling reports of early anthropologists such as Jane Belo in the South Pacific who could find no contraceptive use among women who yet had children only by their *own* man.

In his second volume of *Science and Civilization in China,* Joseph Needham discusses the connection between Taoist sex and primitive tribal mating estivals. Ritualized sex in both primitive societies and in Taoism come from entirely different roots than sexual activities in Western culture where the emphasis has been on procreation. In the latter culture, male ejaculation is of great importance as it is tied in with fertility and the male ego. In ritualized sex, however, the main concern is "dual cultivation" and bonding within the group and with nature. None of these functions needs ejaculation to succeed hence male ejaculation becomes unimportant. This completely eliminates some of the most male emotional problems due to such things which Western culture labels as "premature ejaculation" and "impotence" – two categories actually created by an overemphasis on male ejaculation. Freed from the constant "sex in the head" preoccupation with ejaculation, the *jade peak,* as the Chinese call the male organ, naturally acts as it was designed by nature to act unless there is a physical disability. In ritualized sex, which is not confined to the genital area, the entire body and the brain receive repetitive stimuli over a considerable period of time. This leads to a condition called "central nervous system tuning," which can be explained physiologically.

The organs of the body are homeostatically interconnected by the nervous system as well as by the brain. The autonomic nervous system (ANS) consists of both the parasympathetic (PNS) and the sympathetic (SNS). The entire SNS can be excited by stimulation of only a few nerves; thus readying the muscle structure and stopping or reducing activity in organs, such as the digestive system, not immediately needed for escape or fighting. The ergotropic (energy expending) response when the SNS is stimulated, resulting in increased muscle tone, excitation of the cerebral cortex and desynchronized cortical rhythms. The parasympathetic nervous system (PNS) responds only to more generalized stimulation and results in pleasurable states such as sleep, digestion, relaxation and, among animals, grooming activities.

Generally speaking, if one of these systems, either PNS or SNS, is stimulated, the other system is inhibited. Tuning occurs, however, when there is such strong, prolonged activation of one system that it becomes supersaturated and spills over into the other system and, in turn, it becomes activated. If stimulated long enough the next stage of tuning is reached where the simultaneous strong discharge of both autonomic systems creates a state of stimulation of the median forebrain bundle, generating not only pleasurable sensations but, in especially profound cases, a sense of union or oneness with all those present. This stage of tuning permits right hemisphere dominance thus solving problems deemed insoluble by the rational hemisphere. Furthermore the strong rhythm of repetitive action as done in sexual rituals produces positive limbic (animal brain) discharge, resulting in increased social cohesion; thus contributing to the success of such rituals as bonding mechanisms.

From their research into ritual actions, D'Aquili and Laughlin conclude that when this "tuning" occurs in rituals, "It powerfully relieves man's existential anxiety, and, at its more powerful, relieves him of the fear of death and places him in harmony with the universe... Indeed, ritual behavior is one of the few mechanisms at man's disposal that can possibly solve the ultimate problems and paradoxes of human existence." All these benefits which follow on "tuning" apply even more powerfully in ritualized sex; such "tuning," however, cannot come about through the usual quick orgasm of Western style sex, programmed as it is for procreation only. Such an orgasm resolves only the immediate sexual tensions. Ritualized sex requires considerable time to allow for full "tuning" of all the interconnected systems in the body. Jolan Chang quotes Li T'ung Hsüan of the seventh century in his commentary on the "thousand loving thrusts" of the "jade peak": "Deep and shallow, slow and swift, direct and slanting thrusts, are by no means all uniform... A slow thrust should resemble the jerking motion of a carp toying with the hook; a swift thrust that of the flight of the birds against the wind." From his own experience, this modern Chinese, Jolan Chang, adds in a footnote that the "thousand thrusts" can easily be done in half an hour at a very slow rhythm.

Taoist sexual rituals sensitize the entire body; whereas, in Western culture we have forced most of the passion of living into the narrowness of genital sexuality. But any time there is a total response in any situation the whole being is there and, because the whole being is sexual, sexuality is always there in any total response. It can occur in any relationship — with an animal, with a flower, with the world itself. Some of the criticism of D.H. Lawrence came from his total response to nature. Walking one day with his stuffy Edwardian type editor, Ford Madox Ford, they were talking of literary matters; when, suddenly, according to Ford, Lawrence went "temporarily insane." He knelt down and tenderly touched the petals of a common flower and went into an "almost super-sex-passionate delight." Ford admitted that this was "too disturbing" for him.

While written accounts of Taoist sexual practices date back to over a thousand years ago, when it comes to primitive ritualized sex, we have no written material. Such cultures were not literate and, in the case of the American Indian, by the time these cultures became literate their cultures were largely destroyed. Of course anthropologists have written on these matters but they can deal only with surface manifestations. Fortunately because of a most unusual combination of personal talents and circumstances in his life, the linguist, Jamie de Angulo has contributed some insight into the Indians' own understanding of the place of sex in their world. De Angulo's life was so unusual and controversial that no biography has yet been written; therefore a few details of his life are necessary to understand why he was able to transmit unusual insights into primitive sexual matters.

Jaime de Angulo was born in Paris in 1887 of Spanish aristocratic parents. At eighteen he left for the United States where he worked as a cowboy on ranches in Colorado, Wyoming and California. Talked into getting an education by a Frenchman that he met in California, he enrolled in the now defunct Cooper Medical College in San Francisco and, later, got his M.D. from Johns Hopkins in 1912. About this time he married Cary Fink, who later left him to study with Jung in Zurich, eventually becoming one of the major translators of Jung's work. During World War I, Jaime was a doctor and a psychiatrist for the early air corps. He later met Jung and visited him in Switzerland.

It was while Jaime worked on a ranch in California that he became friends with Achumawi ranch hands; thus beginning his linguistic studies. Later he lived with the Pit River Indians. In 1915 he drove a herd of cattle five hundred miles south to the Big Sur country to homestead. There he met Kroeber and Radin, both outstanding anthropologists at the time. Recognizing his linguistic talents, Kroeber got him to teach two courses at Berkeley: one in psychiatry and the other on the mind of the primitive man. Jaime did not stay long because he was not academically inclined. He learned seventeen new Indian languages in the next fifteen years and became known as something of a shaman. During the nineteen forties, when he was quite old, Jaime gave a number of radio talks about Indian culture. Indian people would call in and say "Jaime's ear for song and custom is very, very accurate."

The combination of his shamanic knowledge and his academic knowledge of psychiatry and medicine along with his incredible linguistic ability gave him access to areas of Indian thought denied to other white men. The poet, Robert Duncan was de Angulo's typist during the last year of his life while Jaime was dying of cancer and making a last effort to organize his manuscripts. In an interview with Bob Callahan of Turtle Island Press, Duncan explains a few aspects of what Jaime had discovered. Duncan reports that Jaime got the idea from the Indians "that you could cross over not just between the living and the dead, as in Shamanism, but also from one sex to another." He found the Indian understanding reasonable enough. We get confused about something like homosexuality. But sex and gender are not the same thing... At one point when I confused sex and gender, Jaime said "you're Western in your thinking, you think that male and female are genders. For Indians there can be five genders in a language, five genders in a tribe."

In English we have only three genders: male, female and neuter. According to a number of sources there can be as many as eleven to fifteen genders in some tribes. The most commonly known example of crossing from one sex to another is the "*contraire*" in certain Plains Indian tribes. Influenced by "Thunder," this man could marry a warrior and was considered a sacred personage.

Coming back to the interview, Bob Callahan remarked: "With de Angulo, almost for the first time, we are getting a text on Native America which is sexually charged, we are getting an American Indian mythological text which is sexually sophisticated in terms of the individual identity of the characters. I mean it was there all along, but now we got this unusual man picking right up on it." Duncan replied, "Oh yes. Again, Jaime's definition of sex is not specifically male or female. It's not locked to gender in that way. It's who you can actually fuck. Now that might be a tree, a rock... And it may not have anything to do with producing children... We've confused generation with reproduction and production; and so, as we also have private property thrown in there as well, we confuse children with "our" children. And we can't understand a social structure where the tribe as a whole has children... We still have the term generation, 'for my generation', my generation was the 'jazz' generation. In that way we understand what generation is. The total society generates, and the children belong to that society... Everybody really exists in a continuous world of generations, of being the children of the world they are living in, so you're a very bad child indeed if you do not venerate Father Tree, or any other aspect of your parent world."

The word, engendering means, according to the dictionary, "to produce, give existence to living beings." Looked at from Jaime's point of view, it is obvious that humanity alone cannot engender children – instead it is the entire living environment which produces the child and keeps it alive – the air, soil, plants and animals of its immediate environment. We are the children of our particular place on earth. This is why the land is sacred and sex is sacred and eating is sacred; because they are all parts of the same energy flow as the Tukano and the Taoists conceived it. The Indians repeatedly acknowledge "all our relatives" in their sacred sweat lodge ceremonies – the hot rocks, the water which is thrown on the rocks, the sage – all are part of the same family. Other cultures have never lost this understanding of being part of the whole, the Tao; Western culture did forget but now we have been forced by ecological disasters on every side to begin to re-cognize the inter-relatedness of all:

> *The last three thousand years of mankind have been an excursion into ideals, bodilessness, and tragedy and now the excursion is over... it is a question, practically, of relationship. We* must *get back into relation, vivid and nourishing relation to*

the cosmos... The way is through daily ritual, and the reawakening. We must once more practice the ritual of dawn and noon and sunset, the ritual of kindling fire and pouring water, the ritual of the first breath, and the last... We must return to the way of "knowing in terms of togetherness"...the togetherness of the body, the sex, the emotions, the passions, with the earth and sun and stars."

– *D.H. Lawrence*

REFERENCES

Brand, Stewart. "'For God's Sake, Margaret,' Conversation with Gregory Bateson and Margaret Mead." *Coevolution Quarterly*, Summer 1976, pp. 32-44.

Bureau of American Ethnology Reports: vol. 9 (1887-1888), vol. 26 (1904-1905), vol. 27 (1905-1906), vol. 44 (1926-1927) and vol. 46 (1928-1929).

Chang, Jolan. 1977. *The Tao of Love and Sex*. New York: E.P. Dutton.

Cohen, Mark Nathan. 1977. *The Food Crisis in Prehistory*. New Haven: Yale University Press.

Coon, Carleton S. 1971. *The Hunting Peoples*. Boston: Little Brown & Co.

d'Aquili, Eugene; Laughlin, Charles D., Jr.; and McManus, John., eds. 1979. *The Spectrum of Ritual: A Biogenetic Structural Analysis*. New York: Columbia University Press.

Dasmann, Ray F. "Toward a Dynamic Balance of Man and Nature." *Ecologist* 6, January 1976.

de Bary, Wm. Theodore; Chan, Wing-tsit; and Watson, Burton. 1960. *Sources of Chinese Tradition*. New York: Columbia University Press.

Emmit, Robert. 1954. *The Last War Trail*. Norman: University of Oklahoma Press.

Freuchen, Peter. 1974. In Stanley Diamond, *In Search of the Primitive*. Rutgers: Rutgers University Press.

Gould, Stephen Jay. "Our Greatest Evolutionary Step." *Natural History*, June/July, 1979.

Huang, Wen-Shan. 1973. *Fundamentals of Tai Chi Chuan*. Hong Kong: South Sky Book Co.

Hunt, V. et. al. "A Study of Structural Integration from Neuromuscular, Energy Field and Emotional Approaches." Boulder: Rolf Institute for Structural Integration.

Izutsu, Toshihiko. "The Absolute and the Perfect Man in Taoism." *Eranos Yearbook*, 36 (1967).

– – –. "The Temporal and A-Temporal Dimensions of Reality in Confucian Metaphysics." *Eranos Yearbook*, 32 (1974).

Laszlo, Ervin. 1973. *The Systems View of the World*. New York: George Braziller.

Lawrence, D.H. 1968. "A Propos of Lady Chatterley's Lover." In *Phoenix II: Uncollected, Unpublished, and Other Prose Works by D.H. Lawrence*, edited by Warren Roberts and Harry T. Moore. New York: The Viking Press.

– – –. 1960. *Women in Love*. New York: The Viking Press.

Lee, R.B. and Devore, I., eds. 1968. *Man the Hunter*. Chicago: Aldine Press.

Kolata, Gina Bari. "!Kung Hunter-Gatherers: Feminism, Diet, and Birth Control." *Science*, September 13, 1974.

Needham, Joseph. 1956. *Science and Civilization in China*, vol. 2. Cambridge: Cambridge University Press.

Neel, James V. "Lessons from a 'Primitive' People." *Science*, November 20, 1970.

Nehls, Edward, ed. 1957. *D.H. Lawrence: A Composite Biography*, vol. 1. Madison: University of Wisconsin Press.

The Netzahualcoyotl News, 1, Summer 1979. Berkeley: Turtle Island Foundation, 1979.

Olmsted, D.L. 1966. *Achumawi Dictionary: University of California Publication in Linguistics*, vol. 45. Berkeley: University of California Press.

Pelletier, Wilfred and Poole, Ted. 1973. *No Foreign Land*. New York: Pantheon Books.

Reichel-Dolmatoff, G. 1978. "Cosmology as Ecological Analysis: A View from the Rain Forest." *Man: Journal of the Royal Anthropological Institute*, vol. 11, no. 3, September 1978.

Schaller, George. 1963. *The Mountain Gorilla: Ecology and Behavior*. Chicago: University of Chicago Press.

Shepard, Paul. 1973. *The Tender Carnivore and the Sacred Game*. New York: Charles Scribners Sons.

Snyder, Gary. 1977. *The Old Ways*. San Francisco: City Lights Books.

Undset, Sigrid. 1939. *Men, Women and Places*. New York: Alfred A. Knopf.

Ancestors

Tom Brown, Jr., and Michael Tobias

Tom Brown grew up in the remote Pine Barrens of New Jersey under the tutelage of Stalking Wolf, the famed Apache Shaman who taught him the secrets of tracking and stalking, and of surviving in the wilderness. His books include The Tracker *and* The Search. *He is widely regarded as one of the foremost naturalists in the world today.*

Michael Tobias: Formerly an Assistant Professor of Environmental Affairs and the Humanities at Darmouth College, with a Ph.D. from the University of California-Santa Cruz.

THE MAN

The Lipan Apache called her Big Misty Mountain. She reared 10,000 feet into rarified stands of yellow pine and silver fir. Her summit, scattered with blue andesite, reigned threatening over all the surrounding desert mesas. A fifty-year old grizzly – last one in the southwest – parolled the immense highland region. The bear and unnameable ghosts.

Big Misty was not the only sacred mountain in the southwest, but one of the last to have escaped the incursions of coal and uranium industry. Indian pilgrims might walk three years and four hundred miles from mountain to mountain, collecting pinches of sacred earth in their medicine bundle for healing purposes.

They called themselves Dineh, the people. Chiricahuas, White Mountain, Jicarilla, Mescaleros, and Stalking Wolf's beleaguered tribe, the

Lipan. The Zuni Indian word for them was *apachu*, and the early rakehell Spaniards pronounced it Apache.

The Dineh branched with the Navajo sometime back, wandering down into western Texas and southeastern New Mexico. They were hunters of javelina, badger, deer; they gathered their staple, maguey plant, superb for fashioning both tequila and rope; they took occasional captive wives—only the pretty ones—from among the Pima and Papago, did exquisite jewelry work and moved surefooted, indomitably. These burgundy-skinned nomads had their eyes fixed on Draco and Cassiopeia, spoke with animals, the trout, coyote, fox, wolverine, eagle, hawk and rattler, which was saying alot! All the while they fed on melon, beans and squash.

In 1968, if you'd asked me who the President was, how much one could expect to pay for a gallon of skim milk, or of gasoline, I wouldn't have known, or cared. I only knew that Stalking Wolf had left. He was in his early 90s, more agile and tricksterian than ever, but he was troubled, and sick of New Jersey. Stalking Wolf feared for humanity in the 20th century. He was a mild thinker, a scrutinizing perceiver, but an overflowing *feeler*, in that order. And he *felt* that modern times weren't going too well; that a disaster—caused by man exclusively—was in the making. I was too young, I suppose, to glean hidden alternatives that may have swam around in his mind. But I know he wanted to get the hell out, to go home.

"You look sad?" I'd said to him our last night. We'd been sustaining an all night vigil over the new moon in the Good Medicine Cabin. Stalking Wolf had tamed a snake. Or the snake had tamed him. In either case, the mark of excitement scintillated all over his rufus body. With animals, Stalking Wolf fell into a look as philosophical as tobacco smoke. His eyes foretold much, giving over the touch of paradox to everything he spied and fingered, and heading for the limits of exposé like the headlights of a barrelling train turned inward, or glowing softly amber, mute and sullen, with the blue blue far away look of antiquity. I felt as if I were in contact with the hearth of hearts, when he spoke. The world was electric that night, Whitmanesque.

Stalking Wolf sat resolutely thoughtful, a living anachronism. There was a sense of his coming-of-age, a doubt that blossomed across his fine-boned expressiveness and spoke of retreat from this very century.

"It's time for me," he cached. "Time to go home."

"Where is that, Grandfather?" A beguilling grin.

"Home? Home is the absence of conflicting desire. With my people."

"How many are your tribe?" I asked.

Stalking Wolf couldn't read or write. He never once picked up a telephone, as far as I know; he'd have no way to know about the well-being of his compatriots other than through intuition. But that was more than sufficient.

"Once there were 200."

His hair was black, braided with red willow bark. His face shone like a newborn, eyes brilliantly lucent. He sucked on his index finger, on the fine hairs of his muscle, and squinted after mystery. He moved with unexplainable speed, like a cheetah. His only vice was rocky road ice cream.

Stalking Wolf had once been married to a Caucasian anthropologist. It may have been an instance of empathy between two cultures. Later, he was divorced, never mentioned her. Spoke instead of a pretty little lady named White Deer.

"Where does she live?"

"Border area," he went on. "Near Big Misty Mountain."

"How old is she?"

He thought for a moment. "Old enough to get afraid sometimes."

"Why?"

"It's like happiness."

"Fear is?"

"Yes."

I was 17. Wiseacre, all. Full of reverence for what he'd given me, but cocky and vain with the knowledge of my separateness from high school peers who went to proms, struggled through *Madame Ovary,* as the book was donned, played football, got recked. I was different. I had this bruise in my soul, the far-away alienation of bearded morbidity come into conflict with the glare of a flashing light: illegal to hitchhike; arrested for nudity. But it's a sacred Delaware swimming hole. You gotta' be naked! *Let's go buddy!* Upset with the draft board, with nearly every facet of society. Mom, Dad, I'm a misfit, what can I say! And they'd linger, not wanting to affirm the possibility that it may be true. And all the while I heard Gregorian chants, clacking gaggles of blue heron. Maybe I was insane. Maybe Stalking Wolf was, too.

I remember precious little of our last conversation. As in the cuneiform of birch bark, there was a honed simplicity that seduced the senses, left them startled for all the suddenness, the acuity of his personality. His words counted for sheer subtletly, like twigs in the larger beaver dam of life, Haiku.

He had given me my blood, raised me on the sanctity of introspection at an age when other children wore braces and worried about getting A's in penmanship. I was lucky. But now his tutelage was over. I had to make a commitment if the learning was to be continued in any sense.

"What will you do?" I choked. He looked down. "No!" I cried. "I don't want you to go. Please!"

I didn't know if I could live without him.

"I must exercise," he said wistfully.

Oh he was limber alright. Like a red ant. If I describe his legs as rubber and steel, does it make any sense? He moved like water, like slow, mashed potato cumuli across a Kansas summer sky. He flitted like the chickadee, or crept as a lynx. Exercise? He knew no regimen in life, none I'd ever detected. But at 90 he could outrun me. Stalking Wolf offered the prayer of nonchalant sovereignty, walked deliberately and calmly through his years, mellow, prepared to tense on a lichen, to stalk a mountain lion, but ever untensed. His eyes were cautious, aflame with a love I strove to meet, love of the land. And his words grow together now, as humus from way up high, vert, spread out under the dome of heaven, a vast forest of happy, holy enunciations; late at night secret biddings between lovers. Story line? His read like Nature's itself, risking cliché in this age of wilderness popularism. There has not been such nature fever, such ambiguity and promiscuousness and imposition, since the time of the Romans. But the Apache stand out, for me; and especially Stalking Wolf. His life looked like a mountain farmer's in the Caucasus – anonymous, long-lived, pure; punctuated by a tree slamming down once in twenty years, squirrels mongering their hordes with chattering debate, hawks swooping through the morning motes that pierced the twisted catalpa glens, and the efflorescent medley of tunes, of springs bubbling, storms blasting, rain pattering. I enlarge him, am swayed by the gravity of his urgings and soul. But you see, he was no ordinary Apache medicine man. He was different, ate meat, fish, travelled from tribe to tribe forging a pantheistic spirituality that encompassed many religions. His was a large and tolerant harmony, discrete, ever unannounced, but bristling with the same light-heeled elegance and breadth of true feeling as a John Muir.

Stalking Wolf's only offspring, a son by the anthropologist, had no understanding of the father he rarely saw. Stalking Wolf made few friends of human beings in his life, wandering instead from one wilderness to another, nearly for a century, befriending animals, plants. Lonely? With the Milky Way overhead? Such was Thoreau's response to claims he must have gone nuts out there all alone. Stalking Wolf looked closer to home for his comforts and in such respect was never lacking for company. Of course Thoreau was no total solitaire, either. He took walks nearly every day into Lincoln or Concord, sat around the Central Store, picked up a few cents now and then selling his beloved pencils, even lectured at Harvard once or twice, and went to Sunday supper at his Mom's. Stalking Wolf had his own alter-ego, or so I was able to glean from his numerous mentions of the lascivious, if stellar, relationship he'd had with White Deer.

"You'll be alright," he said, matter-of-factly. The night passed. And in the morning he was gone.

I don't know how he managed, by Greyhound I guess. I can see him sidling up to a little girl on the bus, beckoning with that Stein Erickson kind of smile, big white buckteeth – no cavities which he attributed to his deft manner of rubbing them each day with salt – and that burnished unmarred face; any child would have the same open-door curiosity I had had. And Stalking Wolf, in his long buffalo skin – so out of place in Detroit – would giggle, tease, give away every secret under heaven.

I too went away. My parents would have preferred my taking a job, or applying to Rutgers. I had other ideas, of course, and commenced my own odyssey across America, back and forth more than once. Nothing as fancy as the stuff in National Geographic, but solid bushwacking. I ended up in Maine, fasting for weeks. When a vision of Grandfather, as I called him, grabbed me, sent me in disturbing flight after him. I'd seen a medicine bundle, and remembered the graphic surroundings. But where was the place, actually? A cliff, and buzzards bending over it. The spastic thorny legs of a larger ocotillo bush twisted west. Shadowed escarpments in the inferno of desert upland. I had an idea, only.

But beyond it, a mere clairvoyant urging for such and such a quadrant on the map. Stalking Wolf had never been specific. And *Big Misty Mountain* was not like saying central 29 Palms.

Still I felt as if I knew the place. I can't explain this *anti-cognitive* process of assurance but a partial mention of it is in order at this time, prior to describing the events which followed.

The Native American pharmacoepia is extraordinary, comprising thousands of herbal, medicinal, and spiritual plants. Of the third category an Apache shaman would carry yarrow, for example, in a pouch merely for conversation, as he travelled long distances, chewing chia seed for energy. If a pin cactus infected his foot, well then he'd take a square inch of inner red willow bark with his spittle and apply it. Stronger than two tylenol. How do I know? I tried it. And further more I *believe* in it. For a stuffed head Stalking Wolf would take the long, spherical yellow flowers of the mullen leaf, boil them and inhale the vapors. Or drill a hole in the bark of a white cedar, put his mouth against it, and breath. A general antiseptic could be had from boiling acorns or oak bark; for aches and relaxation, tea fashioned from strong catnip and pine; strawberry tea was used for blood purification and urinary problems; deerbone meal mixed with pine tea and mint was taken to correct vitamin deficiency. Which vitamin? What kind of urinary problem? The Indians had answers to these questions. You doubt it? Then did you ever wonder about the exactitude of fitted stones among Inca and Toltec pyramids, or Egyptian ones, for that matter? We attribute a general ignorance to our ancestors and foolishly applaud modern civilizations for having escaped the supposed violence, ill-health and ennui of our forebears. For such logic the Indian had no antidotes.

He could ingeniously prescribe wood sorel or the extraction of tannin from various substances for its remedial workings on a sore eye. A vast scientific nomenclature supported his intuitive grasp of Nature. This doesn't mean that the Native Americans were necessarily scientific, perpetually testing theory over and against reality. Rather, that they believed in the discernible, the overt interactions within the environment, of which they were on self-conscious, equal footing. Belief is important. A Medicine Man in Utah was discovered to have cured many terminal cancer patients. His medicine bundle was carefully examined, its diverse contents subjected to every known botanical and neurophysiological test. The researchers could not decipher any of the plants. The university verified the man's genius. But the AMA shuddered and discredited him.

The Indians believe, foremost, in the pragmatic health of the land, and their worship, their love of nature conveys this genuine quick of steadiness; lends vigor, ruby cheeks, spiritual credibility to the whole collective culture, from Canadian Inuit to Ecuadorian Canari. This health was clearly manifested in the skin tone, the pupil clarity, the aura, agility, mirth and mystery of Stalking Wolf. More than any other quality, this one stands out in my mind – abundant, contagious zeal. For him, life was a wink, a grimace, and always another wink. White man is schizophrenic in his relationship to Nature. For the Indian, who had never

Photo by Richard Carter

heard of Lamarck, or of destroying the very resources which sustained him, science was only so useful as it promoted.

I drove to the Southwest looking for Stalking Wolf and got deeper and deeper into labyrinths of clan ties, rumored enclaves, impassible canyons, mythic lore where the dead and living merged. I got run around, inferences, doubts of shadows, vague vagaries, seated about scrap shanty towns with old demirune Mescaleros, young herders in Ford pick-ups. Finally, I was directed into a high mesa region, obscure, purposeless, save for the fact no one went there. Without road, one of those rare interstices of America that has retained its look and wind-swept scent of wildness you think you sometimes see from an airplane. And maybe you do. But there it was. Desert chinook, holly, scorpion tracks, algae in the selenic potholes. The plateaued world lay open, giving up to the eye, yet inaccessible.

I drove fast, spewing up dust that lingered behind me for two miles. I approached a strange hallowed ground along a river thick with cotton-wood, box alder and fragrant tamarisk. I noted tracks pell-mell across the golden cactus country; the scampering toe marks of a Mexican vole, much like the pack rat. Probably 30 seconds old. And the long leap catches of a blacktailed jackrabbit. Up ahead, two graceful palominos.

Four people were there. Seven wickiups, one with a wiregrass cape-like affair, oblong, its abstemious exterior concealing the fact of lavish inner decor, you could be sure.

A man in his mid-thirties stared hard at me. He held a Winchester. Probably could have sold it to a museum. Three old people sat on woven stools. Two men and a woman. As old as Stalking Wolf.

I turned off the engine. The younger man, seeing a braided, dark youth in loin cloth, mocassins and beaded head scarf, hop unabashed from the Indian's favored vehicle, put down his gun and extended the timid look of wonder that always enamored me of the Indian's manner.

"What?" he said, eyeing the hand-made Bowie knife at my waist.

"Stalking Wolf. He is my Grandfather. I'm looking for him."

The young Indian forwarded my inquiry which caused some discussion. There were killdeer tracks through camp, paced unselfconsciously, indicating fearlessness. Fifty yards past the encampment was a garden, tenuous combs of maize and broadbean sprouting in the arid sun and umber soil. Good earth here, of chert, sard and gneiss. Pungent, hard. In the distance, mountains, the color of purple locoweed. On a creosote shrub alighted a Mexican jay, its eyes flinging curiousness at us. The old woman may have been 100. I moved up to her. She was emaciated and russet, her bones sinuous as lace, high and conferring. She was not a full-breed. Her eyes echoed blue Bremen. A seductress.

"What you got Grandson?" She looked me over with benign interest and stood up.

And then I saw what I'd been looking at. Stalking Wolf's hutch. I recognized the enlarged hawk. Red-tailed. Surrealistic. Painted across a fancily tanned buckskin on the entrance. My fingers went to confirm the discovery.

"He's been gone two weeks," she said.

GHOSTS

The tracks leading up into the ravine were nearly lost in granular sands. Fresher lizard tracks atop them gave away the tell-tale information: the minute impressions of banded geko, 15 days old; twisted slides of zebra-tailed lizards. The canyon glowed carnelian in the waning afternoon light. I was loping semi-circularly in order to circumvent Stalking Wolf's unguessable trail. He'd deliberately made it difficult. I could gather no indication that he was ailing. He was testing me. White Deer had said this was his final walk. Impossible! I thought, stepping up my pace to the challenge of his trackless tracks.

There was added confusion, a science fiction of markings that the soft suffusion of light exploited. Animal tracks riotously askew. I put my face to the ground, glancing sideways, closing the top eye, then the bottom, to gain perspective on depth, and lighting. Not a clue, save for the peppering of erroneous steps: a three-legged bighorn sheep; spade foot toads that hopped backwards, describing in their wild eliptics a kind of madness; cricket pirouette marks among the green wiregrass that shot up ram-rod from alkali encrusted mud; the ever so faint glint of a back wasp's underside along the luscious sweet grasses. The earth was alive with jasper speckles, the sheen of pyrite, garnet chips, rose and yellow quartz, grape-dense clusters of moth on yucca. Against the lee-side of the canyon were incoherent markings of wild burro dragging a travois. Two weeks ago they'd been by. I rubbed my eyes. A covy of sage thrashers launched wide and brilliant from grottoed mud nests high in the rocks. I stooped, examining a coyote notch: impossible! A pregnant male, walking sideways like a crustacean. What was going on! And the vibrant-hued scat of ring-tailed coati in spirals, as if someone had squeezed chocolate frosting.

There was mistral, Mediterranean-like, blowing off the mountain and bringing with it a contagion of odors. I made for a cave in darkness, gathering plants en route; fashioned a fire from dried bladderweed, a tangle of duff, and lay back against cold stone, smokey sputterings lending scant comfort now. Something was wrong with me.

I touched hoddentin – the pollen of cattail – to my brow, then slept.

Fidgeting in the night. There were squeakings across the cave floor.

Sounded like western harvest mice, the kind that do Fred Astairs on your face. I felt for them with my hand. Something wobbled onto it. By the glow of embers I made out dozens of brown bats writhing. Some clung to one another on the seeping ochre walls. Two were copulating, the male hanging from the female, flapping in the dank air. I was nauseous and fled to the entrance. Something, something rushed over me. I was in a cold sweat, standing in the nudity of night. Blasts of wind assaulted me.

I started up the canyon, along the caking grey dirt of a dried river bed. I smelled the ephemeral aromas of Englemann and fir, come down from the Hudsonian life zone high on the mountain. There was Sagittarius, and the nostril-quaking onrush of irradiation.

I moved instinctively. Sharp feldspar chips, the first hint of granite up ahead, bore into the balls of my feet. A ground zephyr picked up. Until I had to fight it to proceed. Plant parts, blown hither, dusted the turmoiled air. Mesquite beans, tumbleweed, iridescent leaves of manzanita, whirring insects in rapid fire hysteria, micaceous slivers of shale blew into me as from across some Gobi. It hurt. I was maddened and fast dug a hole into the coarse gravel. This was the Apache way. When it was deep enough I put my head into the earth and screamed.

"I am kin to all of creation! Fire-hardened! Lay off!"

I was yanked and tossed like spume. I slammed into a yellow hairy man. Pain pierced me all over my skin. I tried to flee. No man was he; but cholla – barrel cactus – six feet tall. The wind hit me again. I rolled. My eyes were weakened against the blizzard. Rocks spalled from the cliffs. Sulphur in the conspiring air. Stains of iron oxide on my forearms and the hint of bat turd.

Suddenly I was rising, not of my own accord. Prone, moving up, I hurtled insults at Ussen, the Apache Creator. A horrible face rushed at me. I stabbed at it with my knife. The implement was ripped from me, whipped into space. I tried to run but reeled back, once, twice, before losing my balance and falling into the creek bed.

I then heard drums through the gusts. Above me, cavorting imperiously, were three dark figures. They wore body paint, and kilts of patched buckskin, and their faces were pitch black, hooded. Eagles, stars, hummingbirds, and monarch butterflies were painted on the wooden slats of the headdresses. They made agitated swipes at me with wooden swords and their eyes burned against the violent darkness. I was engulfed in sagebrush smoke, and the stench of urine.

Clankings, animal screeches from hidden perches. It was the *Gans* attacking. Apache mountain sentinels. Behind them reared the mammoth head and torso of their shaman. He cursed monosyllabic outrage at me. Blue and white triangles were etched into his chest. There was Buddhist bestiality in his angry profile, the wild gesticulating arms, his heaving waist, sexual in its oblivion. It was a Ghost Dance, meant to touch the Otherside. No, it was aggression. They came at me.

My limbs were numb, immobile. I tried to get up but the heavy figures fell on me. I flung them off. The shaman screamed from off his rock and with his sword came whacking down at me. But made of air! For with incendiary suddenness, the four guardians convulsed, then vanished. A hint of phosphorus in the air.

Clouds obscured the night making blackness blacker. I was spinning, ran for my life. Mist descended the uneven ground, enveloping each plant, every rock outcrop. I crouched in an animal squat. Sweat poured from my temples. There was ringing in my ears, and hot juice in my veins, bringing me into an orgasmic kind of reunion with the obfuscated earth. I hurt, I craved for only this.

Lightning! It smacked the throe of mountain. In the unknowable darkness boulders tumbled down all sides of the ravine wherein I grovelled. It started to rain. Harder, until hundreds of runnels vented turgid cavalcades. The river was rising. I sought refuge against the wall. When I heard it! Coming with the velocity of a plane crash in jungle, all in a split final faith second of unimaginable revulsion. Flash flood! My fists jammed into a crack but could gain no purchase there, with all the freezing rain pouring down the fissure. My eyes were squeezed close and pulsing like neon lights. I shook epileptic in the onslaught of elements. Nature's grand importunity brought me to nerve zero.

I turned, facing my destiny head on just as the muddied effluvium overwhelmed me. Time lapse. Gone.

I remember soft impunity, ecstatic. I swam into heaven; I breathed underwater, talking to cowfish. I went to hell and back. A viper nested in my sundered belly and lay its eggs and the eggs cracked and I was reborn, in the damp oval depression along side the loveliest rock I'd ever seen. It was a boulder of turquoise the size of my jeep. Green and black matrices. Dark ultramarine stretches unblemished. A microcosm of simplicity like meadows in which I rested, took stock, breathed calmly, palms to tired eye. A golden eagle in the rent sky. The water had sunk into earth and all the living world was renewed with timidity and trembling.

A coyote mournfully greeted the day from atop the scissured cleft of rock high beyond. The sun had not yet illumined this lone animal. I watched it traverse the rim. Twice it distinctly turned to examine me. I'm sure of it. The certainty broke through prior hallucination, or whatever it was.

Clicking. Pause. Memory.

I looked down. There was the viper from the night. I didn't breathe. It was coiled and rattling. But the head remained unraised. I reached out, took it by the neck and studied it nonchalantly. I don't know why. Another snake came, a hognose. Then a sidewinder. They all slithered up to me to share in my body heat, which was sensible. Stalking Wolf had told me that if you can control the snake, you have achieved a very powerful medicine, though Apaches generally consider them to be demons. Stalking Wolf could thrust his index finger into a biting snake's mouth, get it behind the fangs before it could snap down. This took bullet speed and precision. There was also something of the akido master in Stalking Wolf.

The snakes uncoiled, twisted away. I stood up, dizzied, fell back down. Cacti wounds.

The sun topped the cliffs, splattering gold along the creek where spindly dogwood jutted from the caulky jumble of smooth depressions. Sand verbena, sand sage, saltbush, some cheatgrass spitting up between microdunes.

There was the scent of resin, sulphur flower. A bluebird flitted on a stunted juniper up slope, calling to me. Hot dawn upon the gorge. A four-winged dragonfly fastened itself to Apache plume. There was the fresh dung of a jackrabbit aswarm with gargantuan flies. And atop a patch of sourdock, the bleached bones of a pronghorn antelope, stuffed in a lassitude of transciency – no meaning.

Nothing looked real anymore. I walked half the day, collecting agate, sardonyx, then tossing them. I tasted dockweed, scarlet penstemon, and the bitter Cowania stansburiana, quinine bush. I knew what it was, but I must have been suicidal by that time. There was a tapestry of urgings from the environment, bringing me home. Ravens and magpies kept apace with me overhead. I found the tracks of a weasel, a porcupine, and then of a mountain lion, leading up mountain. No trace of Grandfather.

And yet, somehow, the lion walked like Stalking Wolf! I followed the prints. They were two weeks old. The animal weighed 115 pounds, Stalking Wolf's weight a few years ago. The animal was tired, moving with some pain, a thorn in its left forepaw perhaps. It had arthritis. The right hip admitted irregular stride. I felt for this particular animal. By its walk I could discern extraterritoriality, a sense of being alien, out of her environment, or maybe dying of hunger, sensing demise. Even remotely, in this animal's evocation, I felt the worldly bitterness of Nature's implacable formulas – kill or be killed. Was this it? White mountain gorillas in Zaire ate green plants, no meat. Seventh century Japanese were largely vegetarian, like Charlie Chaplin. But most animals eat smaller animals. We are stricken with conscience, *double intelligence*. Or some people are. I am. Yet I find the scene of a colorful caterpillar writhing under attack of ants, or the fawn pulled down by dogs somehow fascinating, even liberating. If god is this cruel, I say, then I have powers of greater clemency, of a conscience beyond the frontiers of god. Our conscience endows us with love and because of it I have always felt like an island in the sea of evolution; like a freakish protectorate of the senses where the mountain lion and I can lie down together any day of the week and make wuppie.

I stopped. A red ribbon hung from an oldman cactus. Surrounding it, the tracks of a greywhite kit fox. They led to the creek where puddles brimmed with expired minnow. A yellow mud turtle caressed the air on its backside. I righted the poor fellow. It looked at me with an attitude of equal compassion not unlike the boredom we attribute to its kind; then it ambled ever so lunar into the trajection of a river toad. The two avoided contact. Suddenly, a pintail swept down and the toad vanished between clenching jaws. The turtle kept on through a maze of jasper, petrified slate. No thought of its near extinction. Why the toad, and not the turtle? Doubtless the survivor thought not about it.

I followed mesmerized this Socratic bundle of wrinkles and antiquity and slow-motion purpose. On all fours I trailed it.

A tan arrowpoint! For long minutes my fingers toyed with its possibility. Then I stood up, looked around, sensing eyes . . . My heart shuddered. Burial ground! I'd crawled right into the center of one. I stood surrounded by artefacts: sacred stone charms – unworked turquoise chunk, Prussian blue azurite, Kaolin white clay animal figures – coiled basketry, banded calcite, a kiva ringing stone, gourd rattles, fleshing tools, antler flakers, bone knives, hawk heads. There were spherical geodes, shells of abalone, a large slab metate, a pair of turtles and lizards, sacred sage brush, bits of otter fur, snair setups, gnarled sinew. I'd landed on an entire site. And I was trespassing.

I stepped back, turned rapidly, *No!*

Staring face to face with a seated Apache. Bones and skin. A black forearm with blond hairs partially singed. A brass ring on a finger. His mouth retained its last tranquil gesture. I was sick, had defiled this resting place. I withdraw a twist of tobacco and placed it at his side.

"I'm sorry."

I stood up. The morning was thick with bluster. Now the afternoon dug talons in. Something seemed familiar. I'd seen it in Maine, in my vision. There was the same coyote on the ridge. I scrambled up after it, combatting the interminable tallus slope above me. Characteristic of coyote – the Indian's granite master, the trickster, the survivalist – it awaited my ascent curiously, goading me on, then continued to lead me by sudden spurts and halts, to an adjoining ravine. The rock was different now, granite. White, dalmation, fresh as after shave, or Idaho cumulous. There was no way down into the ravine so I followed the ridge along its exposed, long-sinuous edge, drop-offs to either side. The rock shone under moss campion, Sego lily, mountain gentian. I got higher. The sky was lambent, with white strains of blown mist.

The coyote disappeared.

UNITY

I climbed along the escarpment all afternoon. The angle was manageable. I felt free of my earlier visitations, could not even consider them, in the same manner that we can easily dispense with troubling dreams.

I eventually attained the abutment, a plateau surrounded on three sides by vista. Two narrow gulleys intersected some five hundred feet below me. I sat down. On the other side, Big Misty Mountain rose sheer for thousands of feet, clad in jackpine. Suddenly, I was again pursued. A voice, Stalking Wolf's own. I cried out to him.

"Grandfather!"

Shoona, shoona, shoona . . .the oracle trailed off. A drum beat.

I looked backwards from where I'd come. There was the coyote. But it was no coyote. A wolf! The animal approached, instinct with malevolence. I was stunned, back up to the very edge where an orange cliffrose and the saffron flowers and golden stamen of a prickly pear braved the windy rim. The animal closed in, baring its fangs, all prognathic, warring, low to the ground, rabid. It was forcing me closer to the mountain.

I maneuvered the precipitous summital step, dropped over and clung below the other side. The wolf dug in above, snarling, biting down at me. I hastened to a ledge, oblivious to the plumb line of vision beneath. Now the wolf settled on to its haunches, bemused. I looked up exasperated. This was not the behavior of any wolf I'd ever known.

The sun was behind the mountain as I strove downward. The cliff got steeper. A dihedral shallowed out, became incipient and left me standing atop a one foot square pedestal. I was in trouble. Below, ungiving overhangs. I sought out a traverse. There was something. I resolved, forced breath into my lungs, keeping perpendicular to the rock. I spread-eagled along a vein of pegmatite. My knees shook, sewing-machine. As I shifted the weight and momentum over the knee, localized outwards from the hip, a stinging sensation wracked my toes, which were barefoot to the rock. I bit my teeth. The vein was razor-sharp. My toes began to get slippery with blood.

I clung to that damned rock. No luxury for vertigo or contemplation. My hormones had their own urgencies. I worked sloppily across the face, finger by finger, afraid to breathe, to think. Never had my thighs felt so awkward, my toes so limited, my strength as shallow. My chest labored with rasping. If I fell, it was open air, a 200-mile-an-hour arc away from undercut rock, into darkness.My body would burst into gross body parts, a rank, ruinous powder of mucous and brains. I got hot, prayed.

Taking a nodule of quartz in my hand. The rock is cooling by sunset, gives more friction to the fingertips. Earlier, the rock had sweated, greasy as skies in Hoboken. I felt around a blind corner, heard faint murmurings. I lowered myself down to a ledge, off a fist-sized nubbin. There was the toxic delphinium plant, with its deep blue petals, gingerly nestled in a corner, and the aged droppings of cliff swallow. I peered over the edge: a wide chimney descending to the earth, or so I imagined. I'd never put myself in such a place. I fitted my back and rearend against one side, my palms and knees and toes against the other. Then it was one painful down-squirm. My skin was peeled off, but there was no other way. I descended through that gap 200 feet or more.

But the chimney did not reach bottom. I hung in there long minutes, still a drop from the bouldery canyon floor. I couldn't gauge the height, exactly. Twenty, maybe forty feet. No question of jumping. The chimney just ceased, space under me where the rock wall parted. Scratching for anything out on the open face around the corner, holding myself in with my knees. Something beckoned me, then another. But how to get directly under it without my whole body swinging out? A horned lizard, fast to its steep lichen, gawked innocently, bobbing, from a protrusion several feet away. It came closer then changed its mind, repairing up rock. And then something happened which I've never seen: the lizard fell, actually fell off that cliff. I watched it sail free like an Acapulco diver, then hit ground mutely. A stunned pause, and then it raced, fleet-footed, into a clump of hognose cactus. It lived! The fall revealed to me the fact of a shortened distance separating my awkward stance from the lupine-covered earth.

I let go; out in open seas, swinging to the exposed, burgeoning headwall. Then, choiceless, into air altogether, gulping, heart-stoppage, then tumbling downslope. No injuries. I stood up. No injuries.

I looked down where the dust was still settling. An eagle feather. I picked it up, tied it in my hair, and collected my thoughts. My cold training held no special powers for me at that moment. I shivered. The sun had gone down, the air chilled with stippling gusts flushing the arid confines of the canyon. I walked to keep warm, strode unevenly to the base of the mountain, reaching the remains of a campfire. I lay down foetally, without thirst or hunger, covering myself in dirt that had been recently turned up, and passed out.

The sweet sighs of vesper sparrows. Sun. I shook the cold dirt off, stood up, aching in every joint. My face and hands had been bitten in the night, those friendly little desert midge-like creatures that cause such furious itching. Aggregate lumps all over. Something . . .

I turned. Motionless!

Sitting right behind me, sitting for probably 50 years, was a skeleton, and a small drum.

I knelt down beside him, in prayer, took a twist of the tobacco which

I'd grown in the Pine Barrens and placed it on his right side. I stood up to walk away when another twist appeared, this one on the mortal's left side. There was a small depression. I got down and examined it. A knee print. And a toe.

Stalking Wolf!

I gazed on the corpse. Stalking Wolf's own grandfather. The two of them had communed here 19 days ago. Stalking Wolf was 25 when this revered medicine man walked away forever. The white man had nearly wiped out his entire clan, during the period that Geronimo (his real name never spoken) and his band of 39 went into western Chihuahua, having made tentative peace with the Mexicans at Casa Grandes. It was into Mexico that the U.S. cavalry went in order to deliver the uncapturable leader over to General Miles. The Lipan also fled, and Stalking Wolf's few surviving clan members escaped into these blue mountains. The thought, the presence, evinced a single tear, salty and warm.

The tracks were clear now. Stalking Wolf had no more chicanery. He had followed a mustang, then stayed hard to a fox's own dalliance. I detected that the fox and Stalking Wolf had sported with one another, as if playing music in the keen and mutual awareness of the other's harmless presence. Sensory giants. Their tracks led up Big Misty Mountain. From the depth of Stalking Wolf's print, taking into account the deterioration factor of loamy subsurface alkaline soil, with wind, dew, rain, more wind breaching the smooth, salient fracture line, diminishing shadow, I deduced his weight: 115 pounds. Thin. The mountain lion!

Over gray blackbrush, through grama grass, the wind gusted. I climbed all through the day, leaving the imperial sweet of desert clime behind me, until I touched what resembled a glacier lily, on a rock where I'd come to rest. I'm not sure what the plant was. Delicate lavender flowers, alone, trembling like myself. Hard to breathe up there. At times I had sublime vantages of the far-stretched surroundings. A maze of washouts and canyonland engulfing the skyborn peak upon which I trekked.

Stalking Wolf's trail evidenced a quality of fatigue and endurance, in the slouching heels, the direct register. Most Apache medicine men acquired their first power in adolescence, power of the first veil. For four days and nights they'd go out, and by going out they'd be going in, as Muir put it. But this in fact was the trick to be accomplished. A full identification with Ussen. Without water, food, weapon, with only a blanket they would go. A voice came to them from off a bush, an animal turd, a flower, relaying which precise substance to carry for the rest of their days in a buckskin neck pouch. And for the first time in the young man's life, in the shadows of swaying willow, or under the cool cascade of water drops spinning prismatic from the arched roof of a mesolithic cave dwelling, the *hunch* of exodus excites him, his legs go febrile, and he lurches blind over ditches, careens into mayhem across the golden desert floor. This power takes the !Kung aborigine lad into rough backcountry in search of cockroaches by moonlight, gives wonderlust to Kurd shepherds, migratory neurology to cowboys, great confidence to taxi drivers. It's the oldest adrenalin worth discussing, and gave rise to the first adventure stories, which were the *only* stories.

It is the shadow on a mountain that attracts the eye, emboldens the imagination, catapults the appendages. The darkness which hovers alluringly over the center, like an old suspicion of buried treasure. The subconscious understands that black turmoil of stone and cold declevity; seizes upon it with the eagerness of illumination. Clouds pass, the revery gains cumulative terrain, until the full range is exposed to the soul's furthest reaches of desire. A place to take one's first and final walk.

Five hundred feet from a summit. A tall stand of firs. Dark, umbrian soil. Pine needles, moist and glossy, rocking in the wind. Open, on the good earth, Stalking Wolf's medicine bundle.

I stared compulsively at it. Sat down under a nearby tree. Made a bow and drill, a tinder of shavings, got a flame kindled. There was mesquite. I ate it. Then lay awake all night, sustaining a vigil over the bundle, afraid to touch it yet, to even think about what had happened.

I must have dozed off. A dream appeared to me. In it was a great desert. I walked across it. In the very heart of the desert was Stalking Wolf, curled up dead, on his side. His fist was clenched, holding onto something. I tried to pry open the fingers, but could not. His fist was strangely, insistently locked. I closed my own fist, then asked the Great Spirit what was worth holding onto in life. Hope? I wondered. And there was silence. Love? Maybe. The same silence. In this manner I suggested many things that men take to be important in life. But the Great Spirit never answered my queries. My fingers fiddled with the sands. *Earth?* I whispered. And my fist got sweaty.

And there was a deep groan which rose from the earth's bowels until great gulfs spread across the land and everything shook. It was an earthquake which lasted long seconds. I was undaunted, though, sitting perfectly still, at peace, thrilled by the show. Finally the last rumbling subsided. Stalking Wolf's hand had opened slightly. And in it was a clump of sacred earth, still warm.

Morning. I picked up the medicine bundle without a pause and at that instant possessed its contents. Here was the wisdom of the forest, the great omen. A calico bag with bone frog, bone whistles, a tortoise shell rattle, root sticks, blood-stained bits of buckskin, a buffalo cup for sucking out diseases, hairlock, herbs to renew a horse's wind, and old eagle skin.

I took the bag with great emotion, looked one more time towards the summit, where Stalking Wolf's tracks led, and made my descent. I had become Stalking Wolf. He'd given me everything. I felt inadequate, but grateful.

FOUR DAYS OF HELL

In the canyon. Lost. Trying to get out. I came from a family of coyotes, stubborn. Agitated. The bundle had conferred no overwhelming clarity. Had I failed? I wandered for days. The old man was gone.

I was naked now. No knife. Held the bundle in my left hand, and moved randomly down one arroyo after another, directionless. I'd obtained jojoba oil on Big Misty and applied it as an unguent. My body shone in the high desert sun.

The last thing I remember is the pinging of a hawk. My eyes were clouded with sweat. I'd not eaten. The rattle, the strike, happened so fast. On a rocky promontory. I fell over, clawing at the small punctures. The snake was five feet long, and thick. It hung on and I couldn't fling it off. It hung there and deepened its fangs. I reeled, screamed, rolled through dust, vomited.

The horror kept me from mending it. I crawled in circles, prepared to die. I'll never understand why I didn't try, at least. In Death Valley, I'd once been severely bitten. At that time I had been sleeping in a comfortable elevated cave. I knew that rattlers inhabited the same cave. But we'd been leaving each other alone. I'd sleep during the day and forage at night. The summer day heat at ground zero in Death Valley can reach 170 degrees. At night it may plummet 100 degrees. The snakes hid under boulders during the day, for even the cave was too hot for them. But they'd enter it to digest in the dawn hours. It was during such a dawn that I crawled carelessly to bed. My hand went right onto a sidewinder. It punctured my knuckles. I grabbed it by the neck with my other hand, cut off its head, gutted it. Then I lanced my wound and sucked frantically. For days I was delirious, fighting off the toxins in my blood. But I'd apparently sucked it clean. I awoke four days later, dizzy, salt-mouth, ravenous. I checked my piute traps. They'd all been sprung. So I started a grass fire and ate the remains of the snake which had nearly finished me. That was three years ago.

Now it felt even worse. I grabbed at the dirt and squeezed tears in my eyes which bleared my vision. The sun, like a golden blood-clot, a drill bit, bore deadening core samples in my stomach and head. My fingers were numb within seconds. There was acid in my veins and the odor of decay already about my person. I remembered Death Valley and still the pain was too much for self-saving efforts. I hurled ravings all about, swam into feebleness, until, sinking in the glut of acute pain, it was over. No more sounds. No feeling. Nada. Oblivion after oblivion, it went that way, for days, I think, until one night I woke up refreshed.

Coolness across my face, lips hard as stone. The medicine bundle was in one hand, caked, granite dirt in the other. I'd peed all over myself. I stood up, wobbily, and went searching for water.

And I felt free, at peace with the world. I'd gone down to the underworld, had seen an apparition there. The Apache believe that when you die you enter another place in time beside the place you'd all your life inhabited. Simultaneously the dead live on in their own adjacent manner. I saw a group of horsemen ride into a campsite, dismount, erect a fire and roast a white-lipped peccary. They carried bow and arrows, stone hawks. Their horses were fatigued and the travail marked the whole clan. The women were dark and noble looking. There was an old one, with a deep etched skein of wrinkles and piercing defiant eyes. And a younger one, long-haired, Modigliani cheek bones, fierce eyes, much like the legendary Mangus. His woman was dolorous, obeisant, a beauty. But they were a doomed group. History had expunged them from its merciless record. In their hasty bivouac I recognized the defeated slow-motion madness of final exile. But only a dream.

THE BEGINNING

They said I'd been gone for ten days. I couldn't remember.

"What you got, Grandfather?" White Deer said, stepping from her wickiup. I tried to answer. Anticipating, bristling with intuition, she choked, began to sob and went back into her wickiup. After some time I went in there with her. For most of the day I tried to console her. She stroked my cheeks, in turn, and ran her fingers through my hair the way my mother does.

Later that evening we went into a sweat lodge. All of us. White Deer had taken off her loose-fitting blue cotton skirt and leather blouse, so that she was naked as a fawn in spring. She tossed sweet grass on the hot rocks. My nostrils puckered. Throughout the night we chanted and blessed this life.

In the morning I turned and left their encampment.

River Root

William Everson

William Everson has been known for 20 years as one of the foremost Catholic poets of our time. His work is recognized for its ecological sensitivity. Since 1971 he has been a poet in residence at Kresge College, University of California, Santa Cruz.

River-root: as even under high drifts, those fierce wind-grappled cuts
of the Rockies,
One listening will hear, far down below, the softest seepage, a new
melt, a faint draining,
And know for certain that this is the tip, this, though the leastest
trace,
Is indeed the uttermost inch of the River.

Or on cloud-huddled days up there shut in white denseness,
Where peaks in that blindness call back and forth each to the other,
Skim but a finger along a twig, slick off the moist,
A mere dampness the cloud has left, a vague wetness.
But still you know this too is a taking, this too can be sea,
The active element, pure inception, the residual root of the River.

Place a hand under moss, brush back a fern, turn over a stone, scoop
out a hollow –
Is there already, the merest wet, the least moistness, and is enough –
No more than this is needful for source,
So much is a start, such too makes up the rise of the River.

Even this, even these, of little more, of nothing less,
Of each, of all, drop and by drop, the very coolness priming the wind
Alone suffices: this in itself, for all its slightness, can birth the River.

And hence such wetness gains liquid body and cups a spring,

Originally published as: River-Root – A Syzygy for the Bicentennial of These States by Oyez Press of Berkeley, 1976.

Lipped down from a crevice, some stone-slotted vein of the mauled mountain,
A jet of liberation, and in so much is swiftly away.
And the spurt makes a trickle, channelling out an edge for itself,
Forming a bed of itself as it goes, a bottom of gravel.

Two join together, they find a third, the fourth sucks in making a fifth.
One and by one, down crick, over bar, under bush, beyond bend,
They merge and they melt, they start and they stretch.
The frozen glaciers fuse and further, the long high levels give up their gifts.

And now over all the rock-walls the River sweeps, he stoops and plunges.
He has found his scope and is on his way.
Let slopes drop slides, let ponderosas topple athwart him –
Log jams of winter, storm-sundered roots and the breakage of forests
Clog up canyons – for him these are nothing.

He has found his strength and takes no defection.
He has smelled his term in his prime beginning and will not be fended.
He carries sea in his gut: heels in the peaks but his throat at the Gulf:
Spending is all he knows.

Spending, to spend, his whole libido: to spend is his sex.

For the River is male. He is raking down ridges,
And sucks up mud from alluvial flats, far muck-bottomed valleys.
He drags cold silt a long way, a passion to bring,
Keeps reaching back for what he has left and channelling on.
All head: but nonetheless his roots are restless.
They have need of suckling, the passion to fulfill. In the glut of hunger
He chews down the kneecaps of mountains.

And bringing down to bring on has but one resolve: to deliver.
It is this that makes up his elemental need,
Constitutes his primal ground, the under-aching sex of the River.

For deep in his groin he carries the fore-thrusting phallos of his might
That sucks up a continent, pouring it into the sea.
A passion for elseness lurks in his root. As the father in child-getting
Draws back on his body, the furthermost nerves of his great physique –
Beyond the root phallos and the slumberous reservoirs of his seed,

Far up the tall spinal range of his torso, the mountainous back and the
cloud-hung shoulders,
Above the interlinking neck to the high domed summit, the somnolent
skull,
Those uttermost lakes of the brimming brain – so does the river-
phallos draw on the land.

Out of the fields and forests, out of the cornland and cottonland,
Out of the buttes and measureless prairies, out of high ridges, the
remotest mountains,
Out of the gut, the taut belly and smouldering lava-filled loins of the
continent,
The male god draws, serpentine giant, phallic thrust and vengeance,
The sex-enduring, life-bestowing, father of waters: the River.

Flying over at dusk on a clear day, trending across it
Coiled below, a shimmer of light, sinuous, the quicksilver runner,
Deep-linking nerve of the vast continent, a sleeping snake.
You follow it down as the light fails, massive, majestic,
Thick and inert, recumbent, torpid with sentient power.

Slowly night takes it. When darkness drops on the valley
The River, irridescent, beats on through hot clay,
Its need and its passion dreaming far forward a full thousand miles:
Its head in the uterine sea.

For the strong long River
Leaps to the Gulf, earth-lover, broacher, dredger of female silt and
engorger, sperm-thruster.
And behind all his maleness that mulling might.
The gnawn rockheads jut for peaks, ridgepoles of height
Where fork-lightning splices flicked roots in heaven,
Tall sap-swollen trees thrust juice at the sky,
Murmurous with pollen, their potent musk.

And the high cut crags.
There bighorn ram covers his ewe in a rushing tussle, the loose rock
Swirls under chipping hooves; it falls a thousand feet: when it hits
Fire flashes below.

And the water-delled flats.
The mountain buck springs his start in the doe,
Pine-needled earth rucked under his pitch, the rubbed antlers rattling.

And balsam barrens where the grizzly, sullen, roused to slow joy,
Mauls his fierce woman, crazed with desire.

And lily-pad lakes where the bull moose thrashes his scoop-sweep
    head,
The huge horns flailing. In the throes of his mate-move
Tramples shallows, cattails shatter, the black testes swinging. His love
Dredges up sperm, souse of his juice streaking her belly,
Seed-rush to the womb.

And those everlasting plains where the buffalo bull couples his cow,
Massive, the humped mountainous shoulders, domed primordial hulk
    of his head
Reared skyward, that ponderous love.

And the randy squirrel, the scuttling rabbit,
Lolling coyote, the prancing pronghorn.

Out of the teeming maleness of earth the black River plunges.

And over his length streaked birds dip down, sip water up in their
    parching beaks,
Stagger-winged skimmers: slaked they fly on.
The beating drake, the honking gander, their necks
Arched in splendor, gabbling under the mating moon,
Knife-blade wings in that watery couple
Slashing torn reeds, a thrash of pinions. They tread down the bitten
Half-drowned heads of shy hens, a mighty thunder.
They wade through flat water.

And the fecund fish:
Great pikes in their plunge, each trailing his mate:
Quick trout dartling glib river-shallows.

Deep down under
The snapping turtle sulks in his cutbank hole. Over his head
Bigmouth bass break water for joy. Far back on the bayou
One bull frog swells his organ-note gong; the syllable of desire
Booms over the bog.

And the River runs.

And now at last
It rides by the somnolent night-lying cities of man,

Past pier and past factory, the long-spoked avenues hubbed on center,
Past the outlying suburbs, past wide plantations, capacious farms,
Past roof and room, the chamberment of houses, those sleep-
bequeathing beds
Where consciousness sinks, the brain bowing at last.
It soaks up strength in slumber and in love.

And the man
Sleeps by the woman, husband by wife, blond by dark, their bodies
Given over now to a deepness of sleep,
But the souls adream.

And in that dream
Their limbs touch, the suppleness of woman
Slumbers against the straightness of man. Along her body
His own makes meaning, and out of this straightness
A trending seeks. As if, for music, to see if such a meaning is,
The murmur of sex makes a wakening within:
Deep in her dark his meaning moves.

And she wakens.
For there is a touch, a nudge and provocation, a slow alerting,
And in the alertness a growing tenseness.
It runs through the marrow, and out of that nerving, in ponderous
sleep,
They roll together.

Stuprate, the innominate phallos
Knocks at the door.

And the phallos, knocking,
Finds a slow invitational yielding of entrance,
A let of approval, and is roused: his body's stem,
The root of their love, stemmed out of the male, gropes through the
dark,
The labial embracement.

And now they waken.

It has been between them a night this night of discontent, starving,
and many a day.
They have flanged, split long on sheer misunderstanding, sore at the
heart each for the other,

A soreness swollen, sick unto death, long past containment. For in
their estrangement
Denied the flesh, their need now burgeoning, but the hearts still sore.
So once again they talked it out, finding no concord, and their
differences
Lay like a bane, could only turn in that torment,
Each galled at the heart, edging the other.

This near midnight. Come to nothing they sank toward sleep, fitful,
a disquieted slumber.
Outside, the great River, torpid and vast, lay dredging its dark,
That under-holding strength going south to the sea.
Over the house whorled stars hover, the apex of night peaks and
goes west.
Under the shut moon, halved, the mockingbird
Sings all night his slow evocation.

And at last they sleep. And all the years
Inert on their lives, merging toward some total relation,
Some clasp of the twinned divisions of self, some unitive truth
No consciousness claims.

And strangely in sleep
The days of their childhood rise around them,
Under the forming hand of the mother, the father's
Benevolent kindness or his terrifying wrath,
Their young friendships, school chums known and perhaps forgotten,
Faces swimming the memorial void, to again emerge,
Nights under whippoorwill stars, those sundowns of dusk when all
nature listened.
And the rare mornings, dawn streaking a sky,
The low wind crouching out of the east, its pane-blurring muzzle.

And she remembers the words of her mother,
Unspoken, woman's ancient wisdom, mystery of the moon.

And he remembers the hand of his father, powerful on the plough,
The head laved with white light.

And perhaps they dream how first they met.
A cajun girl not turned sixteen she came out of the Mass,
Down steep church steps on a bright Sunday.
And he stood there, tall blond youth from far up the River,

Pausing outside on the sunlit pavement,
Curious, this strange Latin rite.

And they saw one another.
And so again on another Sunday,
And so again.

And after, then, their first true meeting,
One look of utter recognition. They knew from that what each was to be.
Surely were they set aside, each for the other.

And he met her father, a dark man and old, of deep religion,
An earth man and a man of prayer, who looked at the sky with a look of knowing.
And he saw this man indeed knew God, and he feared this man.

And they married.
He stood there stiff and frowning during that Mass,
But he did stand, and he did agree to respect her religion,
And to raise the kids Catholics.

Nor will he ever forget
The gift that night of her mymen-gift,
Delicately, her eyes down and her lips parted.
But she raised her nightgown over her breasts
Because he turned down the covers.

Getting up in the dark
The mystery of woman's blood sensed on his flesh,
And all about him the musk smell, scent-drift of sex, her evocative presence.

And sensing that blood he thought of the Mass, and he feared.

That was the night, the fiery tingle, his blond
Blazed on her dark. She burned blue: his steel
Struck sharp the rhapsody of love,
Touslings of discovery, wonderment and delight,
Its animal rage, that love,
Sheets kicked off, her bodice broken.

In the dawn of that day she chided him, "Look, you have broken my bodice,"

The dark breast peeping out of the rent, and that started them again.

At the Mass he knelt, tall and aloof, puzzled.
The priest drank the Blood.

She left his side then and took from the priest a Bread.
He watched from his pew; he thought of her dark womanflesh in his
clasp.
She came back from the altar her eyes down, her lips parted.

This troubled him.

He stood up troubled to let her go by,
The myrrh of the altar faintly about her: her evocative presence.

Or later, near honeymoon's end, one hot afternoon
When between them bashfulness was no more,
She called to him softly, low-voiced,
Hardly the time of day for love, but she called him.

And going he went and found her undressed,
Half-naked in haste he took her.
They took – a savagery of splashed fire,
Kicking and clawing, hair wild, her teeth
Clicking little knives through her sobs,
Her shuttering loins, the sucked belly heaving.

And being well worn down from so much of loving, the great gauntlet
of passion,
He was long time acoming, pounding and stabbing.
And she spasmed again, then yet again, and at last he came,
The spent seed jetting out of his groin.

And they two collapsed, nor never moved again till dusk.
The summer rain, falling, plunged through the gloam,
The hot drops chafing the leaves.

Or he dreams the night of her childbirth,
And the small cry of his son.

It was this that convinced him –
No house divided against itself –
And he entered the Church: at the stone-cold font
Knelt and was shrived, baptised with his son.

And kneeling again at the altar rail
Thanked God of His greatness.
Nor was there ever a Sunday found him not at the Mass.

And he prayed.

And the second child was a girl,
And he loved that girl as he loved his life,
But his wife loved the boy.

And another girl and another boy,
The days lengthening now with incremental labor, getting food and
fuel,
Buying meats, nights by the fire,
A prayer by the bed.

And he thought on God and prayed long on God,
And he prospered, for the land was young, and faith sustained him.

And the years swirled.
No children now for some time,
And both were content.

And they gained in wisdom.

For always the River ran by the house,
In dusk or dawn, in night or noon,
The River, running, ran through their lives.

And the mockingbird sang in the rich magnolia.

Ruins Under the Stars

Galway Kinnell

Galway Kinnell: A renowned and incisive nature poet, his recent works include How The Alligator Missed Breakfast *(1982), and* Selected Poems *(1982). He has taught creative writing at New York University.*

1

All day under acrobat
Swallows I have sat, beside ruins
Of a plank house sunk up to its windows
In burdock and raspberry cane,
The roof dropped, the foundation broken in,
Nothing left perfect but axe-marks on the beams.

A paper in a cupboard talks about "Mugwumps,"
In a V-letter a farmboy in the Marines has "tasted battle..."
The apples are pure acid on the tangle of boughs,
The pasture has gone to popple and bush.
Here on this perch of ruins
I listen for the crunch of the porcupines.

2

Overhead the skull-hill rises
Crossed on top by the stunted apple,
Infinitely beyond it, older than love or guilt,
Lie the stars ready to jump and sprinkle out of space.

Every night under those thousand lights
An owl dies, or a snake sloughs its skin,
A man in a dark pasture
Feels a homesickness he does not understand.

3

Sometimes I see them,
The south-going Canada geese,
At evening, coming down
In pink light, over the pond, in great,
Loose, always-dissolving V's –
I go out into the field and listen
To the cold, lonely yelping
Of their tranced bodies in the sky.

4

This morning I watched
Milton Norway's sky-blue Ford
Dragging its ass down the dirt road
On the other side of the valley.

Later, off in the woods
A chainsaw was agonizing across the top of some stump.
A while ago the tracks of a little, snowy,
SAC bomber went crawling across heaven.

What of that little hairstreak
That was flopping and batting about
Deep in the goldenrod –
Did she not know, either, where she was going?

5

Just now I had a funny sensation,
As if some angel, or winged star,
Had been perched nearby.
In the chokecherry bush
There was a twig just ceasing to tremble...

The bats come in place of the swallows.
In the smoking heap of old antiques
The porcupine-crackle starts up again,
The bone-saw, the blood music of our sphere,
And up there the stars rustling and whispering.

There Are Things I Tell to No One

Galway Kinnell

Galway Kinnell: A renowned and incisive nature poet, his recent works include How The Alligator Missed Breakfast *(1982), and* Selected Poems *(1982). He has taught creative writing at New York University.*

1

There are things I tell to no one.
Those close to me might think
I was sad, and try to comfort me, or become sad themselves.
At such times I go off alone, in silence, as if listening for God.

2

I say "God"; I believe,
rather, in a music of grace
that we hear, sometimes, playing to us
from the other side of happiness.
When we hear it, when it flows
through our bodies, it lets us live
these days lighted by their vanity
worshipping – as the other animals do,
who live and die in the spirit
of the end – that backward-spreading
brightness. And it speaks in notes struck
or caressed or blown or plucked
off our own bodies: *remember*
existence already remembers
the flush upon it you will have been,
you who have reached out ahead
and taken up some of the black dust
we become, souvenir
which glitters already in the bones of your hand.

3

Just as the supreme cry
of joy, the cry of orgasm, also has a ghastliness to it,
as though it touched forward
into the chaos where we break apart, so the death-groan
sounding into us from another direction carries us back
to our first world, so that the one
whose mouth acids up with it remembers
how oddly fearless he felt
at first imagining the dead,
at first seeing the grandmother or grandfather sitting only yesterday
on the once cluttered, now sadly tidy porch,
that little boned body drowsing almost unobserved into the
agreement to die.

4

Brothers and sisters;
lovers and children
great mothers and grandfathers
whose love-times have been cut
already into stone; great
grand foetuses spelling
the past again into the flesh's waters:
can you bless – or not curse –
whatever struggles to stay alive
on this planet of struggles?
The nagleria eating the convolutions
from the black pulp of thought,
or the spirochete rotting down
the last temples of Eros, the last god?

then the last cry in the throat
or only dreamed into it
by its threads too wasted to cry
will be but an ardent note
of gratefulness so intense
it disappears into that music
which carries our time on earth away
on the great catafalque
of spine marrowed with god's-flesh,
thighs bruised by the blue flower,
pelvis that makes angels shiver to know down here we mortals make
love with our bones

5

In this spirit
and from this spirit, I have learned to speak
of these things, which once I brooded on in silence,
these wishes to live
and to die
in gratefulness, if in no other virtue.

For when the music sounds,
sometimes, late at night, its faint
clear breath blowing
through the thinning walls of the darkness,
I do not feel sad, I do not miss the future or need to be comforted.

Yes, I want to live forever.
I am like everyone. But when I hear
that breath coming through the walls,
grace-notes blown
out of the wormed-out bones,
music that their memory of blood
plucks from the straitened arteries,
that the hard cock and soaked cunt
caressed from each other
in the holy days of their vanity,
that the two hearts drummed
out of their ribs together,
the hearts that know everything (and even
the little knowledge they can leave
stays, to be the light of this house),

then it is not so difficult
to go out, to turn and face
the spaces which gather into one sound, I know now, the singing
or mortal lives, waves of spent existence
which flow toward, and toward, and on which we flow
and grow drowsy and become fearless again.

Blade of Grass

William Oandasan

William Oandasan stems from Yuki and Pilipino ancestry. His poetry is a blend of American Indian traditional, political protest, contemporary spiritual, and senryu pieces. He carries the old ways forward into the new day – "fusions of dream and reality, red hope."

Blade of grass
                    you
and your numberless kind
turn
      and part like gates
before the push of the
vernal breeze.
                  Blade
your greenness which is lost
in the meadow's vastness
                              ever bright
and still growing rich
                           bends the sunshine.
Blade
      you are the spring robin
on a mossy slab
in an opened field, in a budding tree
                                        on a village green.
You rise from the earth
                              the source and end
in all her shapes and images
                                 and you are
an only one
one that will never be
                           again
and many seeds you have left.
As the sun peaks
you stretch toward the endlessness

and the light is touched.
You
    O blade of grass
are as common as the
sun, moon and earth
                    in a line
and as vital to me.

standing quietly
beside eucalyptus tree
standing quietly

A Sea Psalm

Eric Davis

Eric Davis: A remarkably subtle and insightful San Francisco poet, novelist and Hasidic scholar currently living in Jerusalem. His books include two novels, Flowers of the Moon *and* The Springs of Marah; *and a collection of poetry,* Sangres.

Out of the sea-west as a child I hunted for whys.
I wept and longed with the fog frozen in my lungs
And the beat of the copperous gongs of seals
Out, out far past my sight.
A child is a child forever whirling.
I was a whirling boy.

But then my tongue took fire
And the sea-ox wind said speak o man
And I stuttered with all the liquid tremblings of a man.
Gender has no end or exit. It always aches.
It is inner as the blood where angels abide. Ah!

I told myself: "Though we burn, though the blank dogs
Sniff in the ruined cities and our hands, compressed,
Stretch over glass: fragile as that. Inconceivable."
I told myself, as the worm marched in my bone,
"Angels recline in the heart of wilderness and
Wing, bewitch, sing their green passions."
And I kept silent.

But at last, in the last,
After my Jerusalems were consumed
And my fake coverings ground away –
The waters ran pure once more and the salmon
Sang in the streams of my green veins and
The Angels spoke. Their syntax of summer pears rang:
You. The always You.

The forever You.
No bang announcing fear. But rather hope.
Call it hope.

The reeds and ducks blustered in the
South of their thoughts.
The hope of the always wild.
Like the love-quickness of new pubescence.
O the geraniums the lilies the gentians
Have their own time table.
The ferns know when they thirst.
Everything keeps growing: the rainbow oceans.

For the Last Wolverine

James Dickey

James Dickey: Carolina Professor and Poet in Residence at the University of South Carolina. His recent works include Babel to Byzantium, The Strength of Fields, The Early Motion, *and* Falling, May Day Sermon, and Other Poems. *He is the winner of several major awards including the National Book Award.*

They will soon be down

To one, but he still will be
For a little while   still will be stopping

The flakes in the air with a look,
Surrounding himself with the silence
Of whitening snarls. Let him eat
The last red meal of the condemned

To extinction, tearing the guts

From an elk. Yet that is not enough
For me. I would have him eat

The heart, and, from it, have an idea
Stream into his gnawing head
That he no longer has a thing
To lose, and so can walk

Out into the open, in the full

Pale of the sub-Arctic sun
Where a single spruce tree is dying

Higher and higher. Let him climb it
With all his meanness and strength.
Lord, we have come to the end
Of this kind of vision of heaven,

As the sky breaks open

Its fans around him and shimmers
And into its northern gates he rises

Snarling   complete   in the joy of a weasel
With an elk's horned heart in his stomach
Looking straight into the eternal
Blue, where he hauls his kind. I would have it all

My way; at the top of that tree I place

The New World's last eagle
Hunched in mangy feathers   giving

Up on the theory of flight.
Dear God of the wildness of poetry, let them mate
To the death in the rotten branches,
Let the tree sway and burst into flame

And mingle them, crackling with feathers,

In crownfire. Let something come
Of it   something gigantic   legendary

Rise beyond reason over hills
Of ice   SCREAMING   that it cannot die,
That it has come back, this time
On wings, and will spare no earthly things:

That it will hover, made purely of northern

Lights, at dusk   and fall
On men building roads: will perch

On the moose's horn like a falcon
Riding into battle   into holy war against
Screaming railroad crews: will pull
Whole traplines like fibres from the snow

In the long-jawed night of fur trappers.

But, small, filthy, unwinged,
You will soon be crouching

Alone, with maybe some dim racial notion
Of being the last, but none of how much
Your unnoticed going will mean:

How much the timid poem needs

The mindless explosion of your rage,

The glutton's internal fire   the elk's
Heart in the belly, sprouting wings,

The pact of the "blind swallowing
Thing," with himself, to eat
The world, and not to be driven off it
Until it is gone, even if it takes

Forever, I take you as you are

And make of you what I will,
Skunk-bear, carcajou, bloodthirsty

Non-survivor.

Lord, let me die   but not die

Out.

The Resplendent Quetzal

John Hay

Conservationist John Hay was President of the Cape Cod Museum of Natural History for twenty years. He is an Advisory Council Member of the Mass. Trustees of Reservations and the former Chairman of the Brewster, Mass. Conservation Commission. Some of his books include Nature's Year, The Great Beach, In Defense of Nature, The Run *and* The Spirit of Survival. *In 1970, John Hay was named Conservationist of the Year by the Mass. Wildlife Federation.*

Under the heading *sacredness*, there are several references in the Oxford English Dictionary to real estate and property, which might reflect the religious yearnings of our economic system, but probably has more to do with the rights it holds most dear. By contrast, the sense in which living things were once held sacred and associated with deities and ceremonial observances has been cast aside in favor of the great god exploitation, sometimes graced with the term development, the envy of much of the undeveloped world. It is a many side deity that has a tendency to crush its worshippers. And no compromise has yet been reached between the scale of modern exploitation and the worlds of nature. They are left behind, to survive as best they can. It is an imbalance that may eventually succeed in destroying several million species and much of the planet's integrity. Yet in some parts of the world which can still identify their heritage in the life of the land, or identify themselves with it, there may still be time to act on its behalf.

What of that extraordinary bird the quetzal, sacred to the Maya and the Aztec, still found in some of the highlands of Central America? Its beautiful golden-green plumes were reserved for rulers. In the *Ancient Maya,* Sylvanus Morley describes the headdresses into which their feathers were woven: "The framework of these was probably of wicker, or wood, carved to represent the jaguar, serpent, or bird, or even the heads of some of the Maya gods. These frames were covered with jaguar skins, feather mosaic, and carved jades, and were surmounted by lofty panaches of plumes, a riot of barbaric color, falling down over the shoulders."

The ancient god Quetzalcoatl was the rain god of the Toltecs, a deity associated with peace and plentiful harvests. Among later people he was the god of life, of wind and the morning star. His name is derived from *quetzal* (the Aztec *quetzallo* means precious, or beautiful) plus *coatl,* the Nahuatl word for serpent. On his back, in the culture of the Toltecs, he wore the resplendent plumes of the bird, and he carried a staff in the shape of a serpent. (Alexander Skutch, the naturalist of Costa Rica, has suggested that the union of these two eternal enemies, bird and snake, might be equivalent to making a single deity of God and Satan. The Toltecs, he writes, may have been symbolizing an end to strife and the beginning of peaceful coexistence. In any case, the uniting of disparate elements into an ecosystem, as we might put it, seems like far less spacious a concept.) The word *coatl* also meant twin brother, which added another dimension, combining as the planet Venus did, the twin nature of a star of the morning and of the evening. So Quetzalcoatl came to mean the god for the morning, and the evening god was his twin brother Xolotl. What greater part could a bird play in the religion of a people?

The quetzal is the national bird of Guatemala, as the much abused bald eagle is our own. It still exists in pockets of humid forests from southern Mexico to Panama, some of which are consciously protected, but it is an endangered species, having been persecuted for its feathers, and because of a growing loss of habitat. It is said that the Maya never killed the quetzal for its plumes, but plucked them from the live bird, so that it could grow more; which had less to do with what we call conservation than with a belief that what was sacred could not be violated.

On the assumption that their Spanish conqueror Cortes was a god, the Mexicans sent him splendid presents, including head bands of quetzal feathers and cotton cloth embroidered with them; and a cruel, vengeful, and greedy god he turned out to be. Moctezuma, forced to fill a room full of gold for his benefit, is said to have sadly protested that he would give it all for the feathers of a quetzal. Perhaps that was the point at which money began to degrade the New World environment, or at least take its visibly godlike qualities away from it.

In spite of its precarious status in some other Central American countries, the Resplendent Quetzal, *Pharomachrus Mocino*, appears to be doing well in the forests of Costa Rica. To a northern visitor unused to so much tropical splendor in birds, except for a scarlet tanger, or baltimore oriole on its breeding visits, the sight of such a bird is almost disquieting. The male I saw looked exceptionally tall and regal as it perched quietly near its nesting hole in a tall tree, with a long curving train of plumes hanging down below the tip of its tail feathers. The green feathers on its back were irridescent, like spoils of light, and its crimson breast had all the openly defiant qualities of blood and the reflected rays of the sun. From the wing coverts, the most elegant green feathers curved over its breast in either side. On the head was a crest of brush-like feathers that reminded me of martial helmets. The two very long slender plumes looked like too much of an adornment for a bird that nests in tree cavities to handle. But, in fact, the male, when he takes his turn sitting on the eggs, faces outward, while the plumes are folded over his back, their tops often seen waving gently in the forest airs. "Solomon in all his glory was not arrayed like one of these."

The quetzal is a shining facet of the great civilization of nature where the spirit of human life was once inextricable from birds and flowers, and tall trees rising from their buttressed trunks with branches smothered in bromeliads and epiphytes, a context of growth and sacrifice reaching through intricate shadows toward the sun. In an open clearing at the edge of the forest where the quetzal and its less extravagantly adorned mate were nesting, a bell bird called with a loud, single "bong!," which was less like a bell than the sound of a metal pipe being hit by a hammer. Inside the forest, nightingale thrushes hauntingly sang, like fine instruments being tuned up to some ineffable scale; and the last I saw of the quetzal was a shimmering waterfall of color plunging down off a branch to disappear in the darkness made by endless leaves.

An endangered species such as a whooping crane or a quetzal stands out because it is spectacular, while innumerable other species, down to pupfish and gophers, that are in equal trouble, get only minor attention; but the primary danger is to the identity of the lands that nurture them. The humid forests of this planet are being destroyed so rapidly that an area the size of England is being lost each year. As a result, it is estimated that a million species of plants and animals, a quarter of those now existing, will go into oblivion within the next twenty or thirty years. The rate of extinction is now approaching one an hour.

Costa Rica is one of those rare countries that decided to save some of its vital areas before it was too late. This small country is about the size of West Virginia, or Vermont and New Hampshire combined, and lies between Nicaragua and Panama. It is divided into dry lowlands, wet lowlands, comprising still existing areas of rain forest, medium altitude regions where the cloud forests are located, and high altitude forests with varying degrees of moisture and rainfall, in mountains ranging up to thirteen thousand feet. From the *paramo* of the high peaks, a cold, rocky region of grasses and stunted vegetation, often covered by mist and cloud, down to coral reefs with their swarming tropical fishes, and beaches where sea turtles come up to lay their eggs, there are any number of biological regions of a wonderful diversity. It was this diversity which the park founders were determined to save, starting only twelve years ago, in the face of the deterioration of a major part of their land. In thirty years, Costa Rica has lost fifty percent of its forests.

The Costa Rican government has now established 22 national parks, based on the original plan of setting aside and protecting as many representative ecosystems as possible. With the magnificent mountain park of La Amistad comprising four hundred and eighty thousand acres, the land area in the park system is now eight percent of the country as a whole. By comparison, the United States has only put four percent of its land area in national parks, and there appears to be a conservative trend toward whittling them down. Costa Rica struggles to protect and maintain its parks and reserves under the burden of a nearly bankrupt economy, while outside these areas, what is left of the original forest is gradually destroyed, and the red earth drained of life.

Aside from the parks, seventy percent of the land is under cultivation. Fruit companies, timber contractors and speculators, more often than not from outside the country, plus cattle ranchers, have been turning large areas of Costa Rica into wasteland. The trees, of which there are two thousand known species, and a great many unknowns not yet recorded by science, are cut down, and other vegetation simply bulldozed away, with a consequent loss of the nation's wealth, its invaluable soil. Forage grass is introduced and hordes of cattle moved in. The tropical sun beats down on the now unshaded earth, the rains wash it away, and after a time it is exhausted and can produce no more; with the result that still more land is invaded. Meat is exported for the benefit of "gringos" eating beefburgers and luncheon meat. The people eat rice and beans. Coffee, subject to fluctuating world prices that often have a disastrous effect on the economy, is the chief crop, followed by bananas, meat and tourists. Export revenue from these sources has been greatly outweighed by what Costa Rica has had to pay for its imports, particularly of petroleum. In the face of inflation, a devalued currency, and the current uncertainty as to when economic stress might turn to social unrest, the natural heritage is still in danger. Eight hundred and forty species of birds, ten thousand species of flowering plants, more butterflies than in all of Africa, still face an uncertain future. So does a people who lose their soil.

As it has in most parts of the world, development, out for short term gains, regardless of long term results, often threatens the very existence of the wild, and all its adventurous species. And it is no easy matter for a small country suffering from population pressure and poverty to resist the exploiters. Since development for "emerging" countries is an ideal, it is difficult to defend conservation unless it seems to lead toward economic or social advantage. Conservation as an alternative to development may only look like another form of exploitation if it is seen as taking food from people's mouths or money from their pockets. But a beginning has been made in this small and vulnerable democracy to set aside parts of a living heritage which in the future may be seen as indispensable.

Flying in by light plane to one of the coastal rain forests now included in the national park system, you pass over wrinkled and folded mountain slopes, brown as well as green in the dry season, here and there smoking with fires, the round clumps of spinach-green trees looking like parts of an architectural model. Through intermittent cloud layers and scanty showers, you pass volcanic mountains and fly above the ragged forests in their dark ground, until you see the white brush strokes of waves that mark the Pacific shore. The little plane lands in a grassy clearing, a narrow strip of uneven ground, with high trees roped by vines on either side, while overhead, elegantly cut and plumaged swallow-tailed kites engage in masterful flight, making tight turns and easy sweeps through the heated air.

The rain forest is a self-sufficient, stable system, composed of a bewildering number of precise relationships, millions of years in the making. The annual rainfall comes to at least four meters, and is not a limiting factor in the growth of the innumerable plants, nor is the temperature, balanced within the system. The principle, dynamic factor affecting the forest is light; its nutrients are locked up in the vegetation, and only superficially penetrate into the subsoil. The topsoil has only a thin layer of humus as compared with temperate zones. The leaves continually drop down to the forest floor, instead of seasonally, and they decay fairly rapidly. Nutrients are taken up by a tightly linked network of roots and associated fungi and transmitted directly to the growth of the trees.

This environment is like a great ongoing, self perpetuating wheel, and within that wheel, species after species interact, in dazzling variety. When you feel the wild energy of the forest for the first time, growth and the consumption of growth seem almost overwhelming, but in spite of intermittent bird cries, the barking of monkeys, the crashing of leaves, or gusts of wind from outside, it is a calm, relatively cool region characterized by an underlying silence. Its multitudes of plants and animals seem to attend upon each other in a great game of ordered alliances, where outbursts of ferocity co-exist with an abiding patience.

The jaguar, whose muscular body flows with the grace of living water, chases and kills a deer or a peccary. Other kinds of predators take their time like the strangler vine which germinates on other trees and then envelops their trunk with a network of its own so as to eventually kill off its host, leaving a skeleton to uphold its own great crown of leaves.

Some of the many glittering humming birds depend for nectar on the periodic blooming of the passion flower, whose large, single blossoms suspend not far above the ground. A species of acacia is protected by ants which occupy its hollow thorns and feed on nutritious parts of its leaves. They defend the tree vigorously against predators, and even kill off the seedlings around its base so that they will not compete with it for food and light. Columns of army ants set off periodically on awesomely organized foraging expeditions, like armies of "unalterable law," moving gradually ahead, to devour every insect, lizard, or even nestling birds along its way. As their ranks move ahead they stir up quantities of flying insects and other creatures of the forest floor, which in turn attracts a crowd of birds to dart around and snap them up.

There is a strikingly beautiful damselfly, several inches across, whose gauzy wings are nearly invisible as it skims over the ground, like water surfaces dancing in the light, except for bright yellow or blue spots at their tips. This insect breeds in species of bromeliads that collect small pools of water, and it feeds on the food that spiders collect in their webs, zipping in quickly and lightly to snatch it and speed away.

You follow the ants at a respectful distance, watch the damsel flies, and marvel at the energy with which this world pursues its ancient directions and alliances. Bird calls from one part of the forest to another seem to signal the stations of diversity. Howler monkeys wake you up at four o'clock in the morning with their roaring, by means of which different troops maintain their spacing with respect to each other. Orange-brown spider monkeys travel over the crossed branches of the high trees, and dive through the leafy, open spaces between them with an easy, reckless freedom; and when gaping, human intruders stand below them they shake the branches so as to knock down debris and try to drive them away.

Out beyond the forest edge is the warm Pacific, with a pitiless sun by day firing a long, gray beach where sea turtles move slowly up at night to lay their eggs, often to have them dug up in unguarded areas by local people, or predators such as dogs, pigs, or coati-mundis. What little ones do hatch out are preyed upon by ghost crabs, gulls and frigate birds as they run the gauntlet to open water, where they are threatened by sharks and predatory fish.

At night the dazzling jewels of the Southern Cross climb their vaulted ladders, and the planets shine like glow worms, or bioluminescent organisms in the sea; while some creatures of the deep forest listen cautiously, and the plants stir in that silence sanctioned by the plenitude of space.

The tropical message is inclusion. The rain forest, with its endlessly varied functions and differences in form, is like a great drawing in, a statement as to the total involvement of life. It is uncompromisingly close, to the extent that anyone too much bothered by the presence of insects, or the mere fact of parasitism and predation, might feel oppressed by it. But without it we lose not only its incomparable species but the original measure of shared existence.

The age is full of tourists, like the writer, who can be transported to almost anywhere in a matter of hours. So we are able to pay flying visits to other countries whose habits and languages may still seem strange but which cater to us as members of a world made one through the power of industry. Even parks and reserves, set aside for their vegetation and native animals, can be seen in terms of what we are able to do to rearrange the world environment, in fact, or fantasy. The wildebeest on the Serengeti plains are balanced on the scales with Disneyland. All the same, we have never overcome nature's capacity to return us to her truths, often catastrophic when we deny them, and to judge us on a more universal scale of needs than we recognize.

A procreant ceremony with its guides in the stars still keeps its initiatives in our being. We are its servants and not its masters. We feel this as individuals through love and its denial, folly and humiliation, failure against achievement, as much as through that superior reason we bring to bear on problems of disorder and anarchy. There is more to us than we can be responsible for. We often see ourselves as hopelessly aberrant and destructive, a nearly uncontrollable mixture of sense and fancy; and yet all this fits the facets of nature herself. The psychic receiver in us, the biological inheritor, still mirrors each new scent or motion in the forest. We have been here before. It is where all our alliances were born, in their infinite degrees of suspension and audacity. The life that can be nourished and consumed at the same time, that grows its beautiful forms in all their means of escape or defense, eternally alert, is still our basic sustenance. That so many original environments are disappearing may have a great deal to do with our culture's refusal to admit a fundamental human equation with them. We dominate them out of existence, and both sides are isolated.

There is, and it has been many times repeated, a limit to industrial growth as measured by available resources; but there is also a limit to our ability to detach ourselves from the other worlds of life and survive at the same time as civilized societies. The lasting examples of shared growth and discovery are not derived from computers. The future of a country like Costa Rica ought to lie with people whose land is known in them, equated with their own lives. Otherwise, more scavengers and exploiters who do not feel it as they do will come in and play havoc, only adding to their poverty. These are the people who know that water, trees and soil measure the terms of their own existence, who have seen them and their wild creatures disappear and have found the world more empty as a result. There is another limit to beware of, beyond which life is no longer held inviolate. If nothing is sacred, then nothing is safe.

As the quetzal still exists, along with the jaguar, the hummingbird and the passion flower, so does the enduring mystery of their engagement with the sun. The gods are still alive in them, and ready to be honored. That which is held sacred counters oblivion.

SECTION THREE

AWARENESS AND REASON

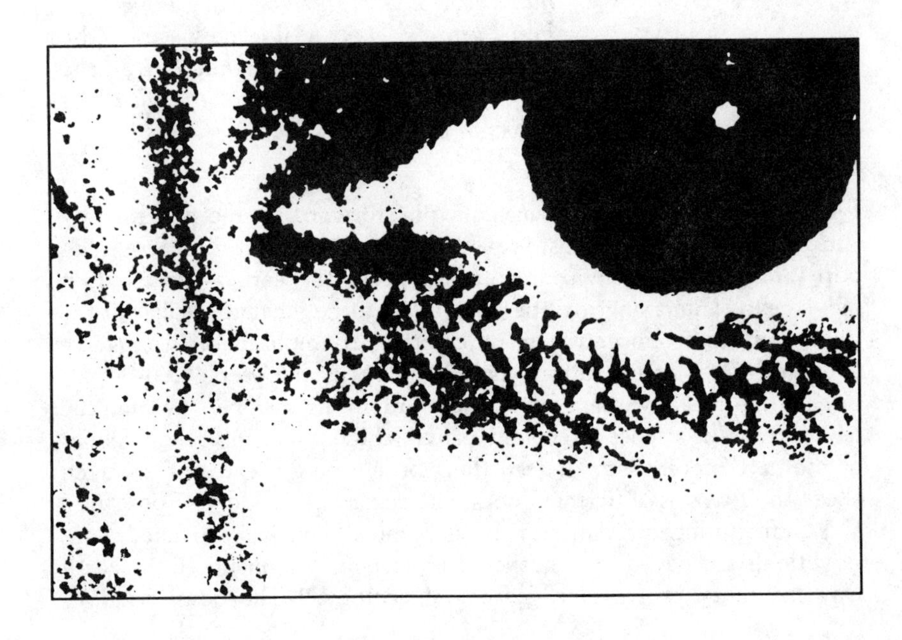

Rounding Out the American Revolution: Ethical Extension and The New Environmentalism

Roderick Nash

A national leader in the field of conservation and environmental management, Roderick Nash has a special interest in problems relating to wilderness and its preservation. At the University of California at Santa Barbara, where he is Professor of History and Environmental Studies, Dr. Nash founded and chaired a multi-disciplinary major called Environmental Studies. Among his eight books and one hundred essays, Professor Nash is perhaps best known for Wilderness and the American Mind, *commended by the publishing industry as one of the fifty best books in print.*

One of the most useful insights put forward in recent years by American environmental historians concerns the qualitative difference of post-World War II "environmentalism" from the earlier conservation movement. Climaxing in the late 1960s, the change replaced the utilitarian, anthropocentric, resource-oriented emphasis of Progressive and New Deal conservation. The integrity of the whole ecosystem, rather than the advantage of its most ambitious member, became the new focus. The science of ecology provided the philosophical guidelines of the new biocentrism. Indeed the word "ecology" explains as much about the 1960s as "efficiency" does with regard to Theodore Roosevelt's and Gifford Pinchot's America. If, as Samuel Hays has explained, conservationists believed in a "gospel of efficiency," exponents of the Earth Day mentality professed a "gospel of ecology."[1] The quasi religious characterization is deliberate; in both periods a movement of political reform acquired the intensity and dimensions of a crusade.

Efficiency had an ally in democracy in energizing conservation in the Progressive era.[2] Again and again Progressive conservationists pointed out that the resources of the nation belonged to all the people and not just, in the vocabulary of the time, to the special interests. This kind of rhetoric, and the idealism behind it, made conservation more than a matter of economics. The wise use of resources for the greatest good of the greatest number was not just prudent but right. Ethics joined economics in making the case to conserve.

W J (he used no periods) McGee, the man Pinchot called the scientific brains of Progressive conservation, illustrates the approach. A follower of the democratic reform philosophies of Lester Frank Ward and Henry George, McGee left no doubt that for him conservation transcended economics. "On its face the Conservation Movement is material," he wrote in 1910, "yet in truth there has never been in all human history a popular movement more firmly grounded in ethics, in the eternal verities, in the dignity of human rights!" McGee went on to explain that the intellectual roots of conservation lay in the idealism of the American Revolution, namely "the new realization that all men are equally entitled to life, liberty, and the pursuit of happiness." It seemed clear to him that "Conservation" (he consistently, and significantly, capitalized the word) had as its most basic goal perfection of "the concept and the movement started among the Colonists one hundred and forty years ago – to round out the American Revolution."[3] Granted that in 1910 McGee was writing about the rights of people, not of nature, but his formulation suggests that ethics could be an integral and explosive part of conservation ideology. It is, in fact, arguable that what gave conservation its special hold on the Progressive mind was this injection of the rights into what ordinarily involved only economics.

The mind of W J McGee offers a clue to the problem of what energized, even radicalized, a later generation of Americans concerned with environmental protection. For the 1960s the compelling idea was that not just people but nature itself has rights which must be respected. If natural rights theory impelled McGee and his colleagues, the idea that characterized modern environmentalism was the rights of nature. For the first time meaningful numbers of Americans could at least think of non-human life forms and even of the non-living environment (rocks, rivers, mountains) as defendable on ethical and not just on economic (anthropocentric) grounds. From this perspective the 1960s witnessed the most ambitious attempt yet to, as McGee phrased it, "round out the American Revolution."

Although there were anticipations in the work of Jeremy Bentham, Charles Darwin, John Muir and Albert Schweitzer, an American wildlife ecologist, Aldo Leopold, did the most to call the attention to the possibility of extending ethics beyond person-to-person relations. As early as 1933 Leopold set forth the essentials of an ethical system that embraced nature as well as people.[4] What he called "the land ethic" received fuller expression in 1949 in Leopold's posthumously published book, *A Sand County Almanac*. At the core of his philosophy was the concept of sequential ethics. Figure 1 depicts and, to some extent, amplifies, what he had in mind. Society, Leopold believed, had evolved ethically over time. This meant including even broader categories in its concept of community. Ethics, which Leopold defined as self-imposed restraints on the freedom of individuals, could be thought of as beginning with the family and broadening to take in a nation, race, and finally, all mankind. What made Leopold's ideas so exciting to the 1960s and 1970s was the idea that ethical extension did not necessarily stop with people. Pets and useful animals, such as horses, were generally included in the community concept. What about other animals? Plants? The entire life community? What about ethical attitudes toward the earth itself – the rights of rocks?

There are, to be sure, many philosophical problems with Leopold's

concept of an evolving ethic. Even as ideals his concepts are not always accurate descriptions of historical experience. And some would call him naive in assuming that human beings were moving inexorably in a linear fashion toward more ethical behavior. Perhaps Leopold underestimated the difficulties involved in having ethical relations with non-human life forms, not to mention non-living matter. But what matters for cultural history is that in the 1960s many were prepared to overlook these problems and use the land ethic as a springboard to new belief and to action. Whatever its shortcomings, *A Sand County Almanac* acquired near-biblical status among the new breed of environmentalists who professed a gospel of ecology. Its brisk sale, in several paperback editions, would have astonished its author who in his last letters in 1948 expressed pessimism about the chances of his ideas ever finding acceptance in a culture which persisted in defining conservation in strictly economic terms.

What Leopold did not anticipate in the late 1940s was the emergence of a broad social questioning not only of economic criteria but of the entire fabric of traditional American value. The "counterculture," as it was called in the 1960s, also stressed the idea of community central to both the ecological sciences and to environmental ethics. Indicative of the new viewpoint was the headquote an activist organization called Friends of the Earth used on the cover of its periodical. The words were Robinson Jeffers':

> *"...the greatest beauty is organic wholeness,*
> *the wholeness of life and things,*
> *the divine beauty of the universe. Love that,*
> *not man apart from that..."*

Friends of the Earth took the final phrase "Not Man Apart" for the title of its publication. It provided perfect expression of the concept of an extended ethical community that included "life and things" as well as people. Earth Day (April 22, 1970 was the first) and the environmentalists' omnipresent symbol of a circle were, obviously, other manifestations of these biocentric attitudes.

Evidence of interest in environmental ethics is widespread in the 1960s and, increasingly, in the 1970s. Professional conferences, a sure if sometimes tiresome sign of the maturation of a field, occurred. One of the first was convened in 1974 at Claremont, California by John Rodman under the title "Conference on Non-Human Rights." Historians and philosophers joined biologists and theologians in discussing the implications of the new ideas.[5] Rodman, a political scientist, had previously argued that liberalism would perish (become static and conservative) if it failed to find new ways to extend the sphere of rights. Furthering human rights represented one frontier, but Rodman was concerned with the possibility of going beyond to support the rights on non-human life forms.[6] In 1980, on the occasion of the tenth anniversary of Earth Day, the University of Denver sponsored a colloquium, "On the Humanities and Ecological Consciousness," concerning extended ethics.[7] The University of Georgia will host "Theological Issues in Environmental Ethics" in 1982. A scholarly journal, *Environmental Ethics,* began publication in 1979, and it is possible to obtain a Master's degree from Colorado State University in "Environmental Ethics and Animal Rights."

Books are well on their way to defining the emerging field. In 1971 Van Rensselaer Potter published *Bioethics: Bridge to the Future.* "What we must now face up to," Potter wrote in his preface, "is the fact that human ethics cannot be separated from a realistic understanding of ecology in the broadest sense." The book was dedicated to Aldo Leopold who, Potter noted, "anticipated the extension of ethics to Bioethics." One sign that this was in fact occurring in American culture was the almost unprecedented interest of theologians and students of religion in environmental responsibility. One of the earliest manifestations was Richard Baer's essay, "Land Misuse: A Theological Concern" which appeared in the October 12, 1966 edition of *The Christian Century.* Five years later the Faith-Man-Nature group of the National Council of Churches published *A New Ethic For a New Earth.* In 1972 two titles are representative of the growing interest: Ian G. Barbour's *Earth Might Be Fair: Reflections on Ethics, Religion and Ecology* and Bruce Allsopp's *The Garden Earth: The Case for Ecological Morality.* The next five years saw many additional popular and scholarly contributions to this inquiry. By the mid-1970s Dennis G. Kuby's "Ecology and Religion Newsletter" reached a nationwide network of individuals, churches and organizations organized in a "ministry of ecology." An interesting offshoot of the tendency to emphasize the religious nature of man-environment relations was, according to Linda Graber's *Wilderness As Sacred Space* (1976), something called "geopiety." Graber defined it as a quasi-religious zeal for particular environments, and she discussed its manifestation in the wilderness preservation movement. For Americans who understood wilderness to be "sacred," its protection had little to do with economics or with human recreation. They fought for the preservation of wilderness because, in the ultimate analysis, it was morally right. William O. Douglas' *A Wilderness Bill of Rights* anticipated some of Graber's ideas.

Books and articles concerning environmental ethics had, by 1980, increased to the point where George Sessions, a self-styled ecophilosopher, could publish a bibliographic review essay citing literally hundreds of titles.[8] A sociologist, Bill Devall, concluded that these writings and the social action they inspired constituted a "deep ecology movement."[9] The "shallow" variety it was replacing was, Devall explained, the old economically-oriented conservation. Deep ecologists had revolutionary new attitudes toward the earth and its community of life based upon reverence for the non-human and ethical attitudes toward nature.

The penetration of these ideas into broader segments of American thought and action is what gave the new environmentalism its unprecedented political clout with the National Environmental Policy Act of 1970 being an obvious example. Robert Cahn was a member of the first council NEPA established, and in 1978 he published an investigation into the extent to which Americans had internalized environmental ethics in their day-to-day activities. Cahn's conclusions in *Footprints on the Planet: A Search for an Environmental Ethic* suggested while still far from dominant an ethical attitude toward the land was far more than just an ideal. The same conclusion can be drawn from recent efforts to save seals and whales from commercial exploitation and possible extinction. Ethics, in some instances, can transcend law as some environmental militants have proven at risk to their lives. After invading national waters to stop a Russian whaling operation, Paul Watson, skipper of the *Sea Shepherd*, explained that "it takes outlaws to stop outlaws."[10] Members of an organization called Greenpeace made sympathetic headlines throughout the nation by literally throwing their bodies between baby seals and the clubs of commercial pelt gatherers. No one had been around to do the same with buffalo calves in the 1870s. A century later the conservation movement not only existed but had changed perceptibly from its utilitarian origins.

Love seemed to be a major catalyst of this change. The new environmentalists loved the earth in ways Theodore Roosevelt's generation of conservationists would have found incomprehensible. On May 21, 1979, for example, a letter from Mark Dubois reached the California headquarters of the Corps of Engineers. Dubois, leader of the fight to save the Stanilaus River Canyon from being flooded by a new dam and reservoir, declared that by the time his letter was read he would be chained to a secret cliff on the river's bank. If the reservoir rose, he would drown. Only one friend, sworn to let him die if need be, knew his location. After failing to locate Dubois the Corps opened the gates of its dam and stopped the reservoir's rise. Released by his friend, Dubois insisted that his action was not a power play but a personal expression of love for a river and a non-violent way of protesting a moral wrong. The same sense of moral outrage motivated the environmental guerrillas Edward Abbey described in *The Monkey Wrench Gang* (1975) as they sabotaged the technology that was, in their view, destroying the West. Although partly tongue-in-cheek, Abbey's celebration of radicalism and violence made considerable sense to Americans with ethical motives for protecting the earth.

Documenting historical change is sometimes easier than explaining it. But the historian of the new environmentalism must ask why this unprecedented perspective captured the imagination of many Americans in the 1960s and 1970s. Why, in particular, should environmental ethics spread from a small coterie of philosophers and scientists to blossom into a remarkable broad public concern? And why should the most recent surge of the American conservation movement be the broadest and most intense?

What is intriguing as an explanation is the possibility that an ethical attitude toward nature is the latest in a succession of American concerns for the rights of exploited or oppressed human beings. The new environmentalism simply transcended the limitations of species. This perspective interprets it as an extension of democracy – a rounding out of the American Revolution in ways that would have astonished W J McGee and other anthropocentric Progressive conservationists. They saw their cause as providing equal access to natural resources for all *human* members of the community. The new concepts made nature part of that community. The gospel of ecology, in other words, expanded natural rights into the rights of nature.

Continuing this pattern of explanation, it appears worth exploring the idea that what powered the new environmentalism was its link to revolutionary democratic theory so central to America's beginnings and subsequent history. The assumption is that oppressed and exploited minorities have always found a soft spot in the American heart. "Freedom" and "liberation" have had explosive potential in this culture. When the emotion inherent in these concepts became associated not just with social groups but with nature, the new environmentalism resulted.

While admittedly oversimplified, Figure 2 may help illustrate this explanation. It assumes that, rhetoric aside, the idea of rights is never universal. The magna carta of 1215 actually had a very limited constituency among English nobility. Later, Americn colonists were clearly regarded and treated as second class citizens of the British Empire. This denial of rights, became of course, an important factor in bringing on the war for American independence.

In the 1770s American democratic theory might have given "all men" equal rights on paper, but in practice some were a lot more equal than others. This comes as no surprise to professional historians who have long understood that the democratic theory underlying the American Revolution had severe limitations. Slaves, women, Indians and, to some extent, laborers without property were, in actuality not equal in rights with other members of early American society.

In time, as Figure 2 suggests, these groups caught up with the mainstream and became full members of the ethical and legal community. To some extent violence, or at least social tension, accompanied each expansion of rights. This seems to be explained by the fact that certain segments of society benefitted from the denial of rights to other segments. The practice facilitated exploitation. England, for example, was reluctant to give the American colonists equal rights and representation because doing so would have compromised the economic advantages of colonialism. That reluctance in time deepened into a determination to fight.

The same pattern emerged in the case of America's slaves; and here the parallels with nature are revealing. Abolitionists objected to slavery because it involved ownership of something they considered as having rights and being oppressed. They saw slaves, in other words, as members of their ethical community. Slavery was not just uneconomic or politically unsustainable; it was wrong. The new environmentalists held the same attitude toward the earth (or "land" or "nature"). Ownership and exploitation were again the issues. Nature, in a sense, was enslaved, and Mark Dubois, Paul Watson and Edward Abbey were, in the same sense, abolitonists. Indeed it is tempting to see William Lloyd Garrison's newspaper, *The Liberator*, as having the same relation to abolitionism as *Not Man Apart* does to environmentalism. Harriet Beecher Stowe's *Uncle Tom's Cabin* (1852) and Rachel Carson's *Silent Spring* (1962) are similarly comparable. And the drawings of chained and mutilated slaves that so stirred the consciences of Americans in the 1850s anticipated the photographs of oil soaked birds, clearcut forests and dammed rivers that energized the new environmentalists a century later.

The similarities between abolitionism and environmentalism have not gone unnoticed. Albert Schweitzer noted early in the present century:[11]

> *It was once considered stupid to think that colored men were really human and must be treated humanely. This stupidity has become a truth. Today it is thought an exaggeration to state that a reasonable ethic demands constant consideration for all living things down to the lowliest manifestations of life. The time is coming, however, when people will be amazed that it took so long for mankind to recognize that thoughtless injury to life was incompatible with ethics.*

Aldo Leopold also used slavery – in Odysseus' Greece – as a way of explaining how ethical ideas could change. In *San County Almanac* he pointed out that Odysseus could hang a dozen slave girls on one rope because "the hanging involved no question of propriety. The girls were "property," and their treatment was "a matter of expediency, not of right and wrong."[12] This was still true, Leopold observed, about land, but he wrote in 1948 before an expanding ethical consciousness began to bring about changes. By 1978 Elizabeth Gray could ask, in the title of a book, *Why the Green Nigger?*. Of course even in 1978 there were a great many Americans who regarded a land ethic as totally crazy. But, it is well to recall, a great many in 1859 said the same of John Brown when he tried to start a slave insurrection at Harper's Ferry.

John Brown's act of violence against what he considered morally wrong, his subsequent hanging, and, of course, the Civil War, suggest that ethical expansion is often associated with violence. Granted that the gospel of ecology has not led to war, but the possibility is always present when an issue is defined in terms of right and wrong. Compromise becomes difficult. Part-time slavery is impossible and so is half a dam. Certainly more than previous conservation efforts, the recent environmentalism has the potential of engendering social conflict just as did the civil rights movement and opposition to the Vietnam war.

It is instructive to note that just as in the case of slavery and the movement for women's and Indian's rights, one of the most prominent goals of environmentalism is opening the American legal system to new definitions of oppression. In 1974 Christopher Stone, a lawyer, published *Should Trees Have Standing?: Toward Legal Rights for Natural Objects.* Stone's point was that Americans were beginning to think "The unthinkable" just as they once had in the case of black and female people. The new cause was to give non-human life forms, and even places, "standing" in court. Of course humans had to speak for these litigants, but a Supreme Court Justice, William O. Douglas, ruled that this was entirely right and proper. In a minority (losing) opinion of 1972 concerning a ski development in California's Mineral King Valley, Douglas held that "inanimate objects" such as "valleys, alpine meadows, rivers, lakes...or even air" have a right to exist and should not be excluded from litigation to obtain that right. Douglas even cited Aldo Leopold to the effect that "the land ethic simply enlarges the boundaries of the community to include soils, waters, plants, and animals, or collectively, the land."[13]

Following this line of reasoning, the 1970s witnessed several well-publicized attempts to resist developments that would have destroyed the habitat of an endangered species. A tiny minnow, the snail darter, was the focus of an emotional, and ultimately unsuccessful, attempt to stop a dam on the Little Tennessee River. Concern for a plant, the furbish losewort, held up approval of the Dickey-Lincoln Dam on Maine's St. John River. In Hawaii, a bird, the palila, sued for protection of its only habitat from the inroads of cattle and goats. The Audubon Society represented the bird, but the case was recorded as *Palila vs. Hawaii,* and the bird won. Significantly, all these species were lacking in utilitarian value to man. The movements for their protection appear to have been based on ethical grounds.

In 1973 Congress passed an extraordinary act that embodied the new ethical perspective and may be taken (see Figure 2) to represent a beginning of the institutional extension of rights to non-humans. The Endangered Species Act guaranteed the right to existence of any species threatened with extinction. Although highly, and expectedly, controversial, it has provided legal grounds for hundreds of decisions stopping or altering human plans. According to Joseph Petulla, the act embodies "the legal idea that a listed nonhuman resident of the United States is guaranteed...life and liberty."[14]

Returning to Figure 2, it is striking that the flowering of revolutionary environmentalism followed closely on the heels of action on behalf of oppressed human groups. The civil rights movement immediately preceeded it, and women's liberation efforts (the Equal Rights Amendment, for example) occurred simultaneously. Several scholars have commented on the relationship of women's rights and environmental rights including Annette Kolodny in the significantly-entitled *The Lay of the Land* (1975). Kolodny feels that male exploitation of women and nature have similar roots and expressions in American culture.

We have, in sum, become accustomed to think of the environmental movement and the ecological perspective that lies behind it as hostile to traditional American values, as part of the *counter*culture. Paul Shepard called ecology a "subversive science," and Leo Marx pointed out the incompatibility of American ideals (like growth and individualism) and the ideals of the new environmentalism.[15] Left-wing environmentalists like Murray Bookchin have argued that meaningful protection of the environment depends on wholesale dismantling of American political and economic institutions. But if conservation is defined ethically, it fits quite squarely into the most traditional of all American ideals: the defense of minority rights and the liberation of exploited groups. Perhaps the gospel of ecology should not be seen so much as a revolt against American traditions as an extension and new application of them – as just another rounding out of the American Revolution.

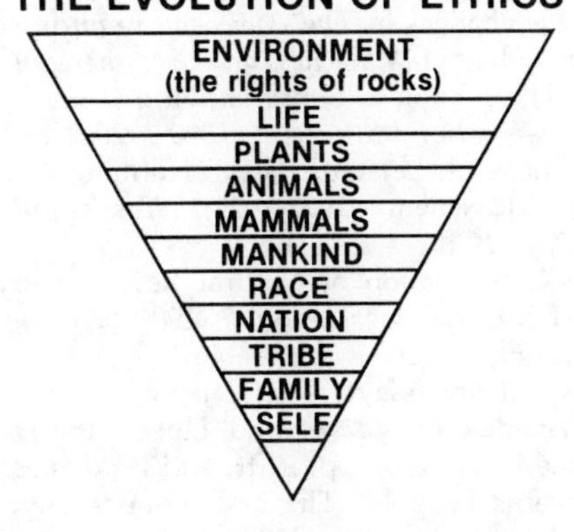

Figure 1

THE EXPANDING AMERICAN CONCEPT OF RIGHTS

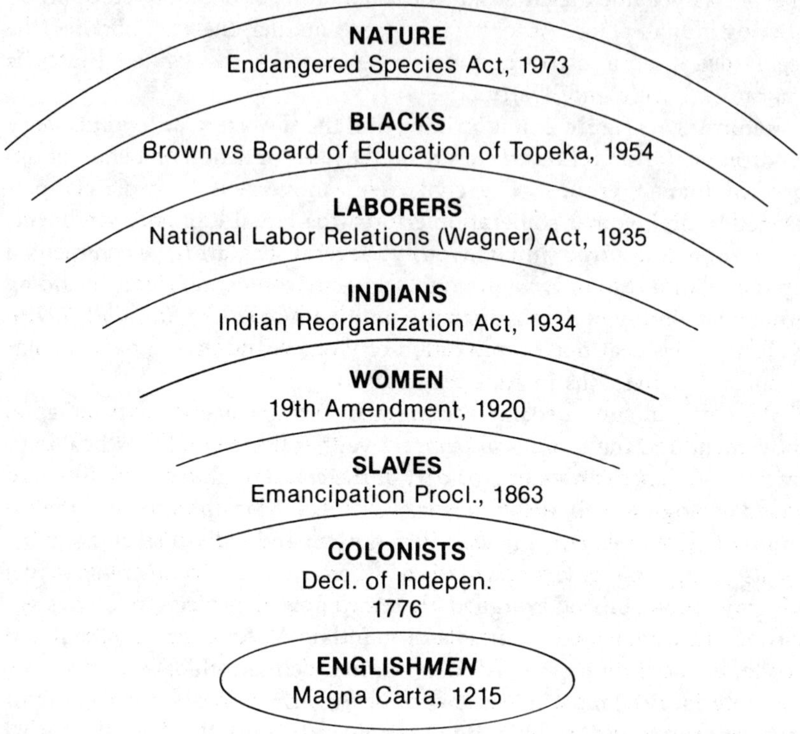

Endangered Species Act, 1973

BLACKS
Brown vs Board of Education of Topeka, 1954

LABORERS
National Labor Relations (Wagner) Act, 1935

INDIANS
Indian Reorganization Act, 1934

WOMEN
19th Amendment, 1920

SLAVES
Emancipation Procl., 1863

COLONISTS
Decl. of Indepen.
1776

ENGLISH*MEN*
Magna Carta, 1215

Figure 2

NOTES

[1]I have sketched these changes in *The American Environment* (Reading, Ma., 1976), esp. p. 226 as well as in *Wilderness and the American Mind* (New Haven, Ct., 1973), ch. 13. Hays' book is *Conservation and the Gospel of Efficiency: The Progressive Conservation Movement, 1890-1920* (Cambridge, Ma., 1959).

[2]The standard reference is J. Leonard Bates, "Fulfilling American Democracy: The Conservation Movement, 1907-1921," *Mississippi Valley Historical Review*, 44 (1957), pp. 29 ff.

[3]W J McGee, "The Conservation of Natural Resources," *Proceedings of the Mississippi Valley, Historical Association, 3 (1909-1910), pp. 365 ff. as quoted Nash, ed., American Environment*, pp. 44-46.

[4]For Leopold see two of my essays: *Dictionary of American Biography* (New York, 1974), Supplement 4, pp. 482-84, and "Elder of the Tribe: Aldo Leopold," *Backpacker*, 27 (June-July, 1978), pp. 21 ff. and *Wilderness and the American Mind*, ch. 11 concerns Leopold. The best monographic treatment is Susan Flader's *Thinking Like a Mountain: Aldo Leopold and the Evolution of an Ecological Attitude Toward Deer, Wolves, and Forests* (Columbia, Mo., 1974).

[5] A related publication was Rodman's "The Dolphin Papers," *North American Review*, 259 (Spring, 1974).

[6] "Liberalism and the Ecological Crisis: Some Tentative Reflections" (unpublished paper presented to the Western Political Science Association, April 4, 1970).

[7] The papers were published as Robert C. Schultz and J. Donald Huges, eds., *Ecological Consciousness* (Washington, D.C., 1981).

[8] "Shallow and Deep Ecology: A Review of the Philosophical Literature" in Schultz and Hughes, eds., *Ecological Consciousness*, pp. 391-462.

[9] "The Deep Ecology Movement," *Natural Resources Journal*, 20 (Spring, 1980), pp. 299-332.

[10] As reported in *Animal Welfare Institute Quarterly*, 30 (Fall, 1981), p. 5. and in Paul Dean, "Soldier With a Dangerous Task: Let Their Be Whales," Los Angeles Times, Feb. 3, 1982, pt. V, pp. 1, 4.

[11] As quoted by Donald Worster in "The Intrinsic Value of Nature," *Environmental Review*, 4 (1980), p. 44.

[12] *A Sand County Almanac* (New York, 1949), p. 201.

[13] Douglas as quoted in Stone, *Should Trees Have Standing?* (Los Altos, Ca., 1974), pp. 73 ff. The quote from Leopold appears on pp. 83-84.

[14] Petulla, "American Institutions and Ecological Ideals," *Science*, 170 (Nov. 27, 1970), pp. 945-952.

[15] Marx, "American Institutions and Ecological Ideals," *Science*, 170 (Nov. 27, 1970), pp. 945-952.

Discriminating Altruisms

Garrett Hardin

Garrett Hardin is Professor Emeritus of Human Ecology at the University of California. His paper Discriminating Altruisms *was prepared while he was with the Environmental Fund in Washington, D.C. Earlier versions of the paper were presented at the American Association of the Advancement of Science in Toronto (4 Jan. 1981); at Southwestern University in Georgetown, Texas (15 Jan. 1981); and at Rollins College in Winter Park, Florida (21 March, 1981).*

According to the *Oxford English Dictionary* the word "altruism" (Latin *alter*=other) was first used in 1853, following the introduction in 1722 of the word "egoism" (Latin *ego*=I). Were people unable to discuss motivation and the consequences of human actions before these nouns were coined? Certainly not: contrasting adjectives ("generous" and "selfish") and their related verbs ("to give" and "to take") sufficed to deal with the contrasting phenomena of social life. But the creation of the nouns – substantives – moved the discussion to another plane by suggesting that there was a thing, a substance as it were, behind each kind of action. In the Indo-European languages (and many others) nouns imply a reality that is greater (more *substantial*) than that suggested by verbs and adjectives.[1] Once a substantive is created it is all too easy to assume a substantial reality behind the word. An unsophisticated public is inclined to put the burden of proof on the iconoclast who doubts the substantive. This stance is 180° wrong. None the less, in the case of the substantive "altruism," biologists have accepted the burden and have shown that, strictly speaking, altruism does not exist; or, to put the matter more exactly, altruism, though it may *exist* discontinuously in space and momentarily in time, cannot *persist*, expand and displace the natural egoism of a species.

Many people find this disturbing news. Fortunately we need not give up "altruism" altogether. We use many colloquial words that are, from a strictly scientific point of view, indefensible. For example we speak of the "cold" of a winter's day (note the substantive), though physicists have convinced us that there is no such *thing* as cold, only degrees of heat. Instead of complaining of the "cold" of −13° Fahrenheit we should speak of the "heat" of +248° Kelvin. But that is pedantry; not even physicists use such language in everyday life. When employed with sufficient care, inexact colloquial expressions do no harm. "Cold" is one such colloquialism, "altruism" (as we shall see) is another.

The sufficient care that we must exercise with "altruism" is this: we must modify the substantive "altruism" with the adjective "discriminating" – or use the noun in such a way that the audience infers the missing modifier. Pure altruism is so rare and unstable that policy need make little allowance for it; but impure forms of altruism – discriminating altruisms – are the very stuff of social life.

THE IMPOSSIBILITY OF PURE ALTRUISM

Before we comfort ourselves with the impure altruisms that can exist and persist, we need to accept this basic fact: *A species composed only of pure altruists is impossible*. The simple theoretical proof of this fundamental principle is found in the following "thought-experiment."

Let us suppose that I am God. I wish to construct a species of animal in which every individual is a pure altruist, i.e., a being that prefers serving others to serving itself. Put another way, when there is a conflict between serving others and serving self, the individual acts in such a way that the benefits of his actions accrue more to others than to himself. Since (by hypothesis) I am God, there is nothing to prevent my creating such a species. *But not even God can make altruism persist.*

Why not? At this point we depart from pure theory to commit ourselves to a single empirical fact, namely the inevitability of random mutations. ("Random means random in terms of the species' need, not in terms of the chemistry of the genetic material.) In the language of the thought-experiment we assume that not even God can put an end to the mutation process. In creating the elements with the properties they have, God committed the living world to change.

Those who like to reinterpret the story of Genesis in the light of new ideas and facts might note that the "firmament," which surely must include the elements of the Periodic Table, was created on the second day, but living things were created later, plants on the third day and animals on the fifth. The Fundamentalists' belief that God's creation was final and incorrigible implies that the dynamic, unstable characteristics of atoms and molecules was inherent from the beginning, leading inescapably to the instability of the genetic code of plants and animals. (This paradox needs to be called to the attention of Fundamentalists who rest their faith on the unchangeability of biological species.)

Once we recognize the inescapable fact of mutability we must acknowledge that the hypothesized pure altruist cannot be what taxonomists call the "type" of any species. Whenever a mutant arises that is less than purely altruistic, the actions of this mutant necessarily benefit its possessor more than the actions of altruists benefit altruists. The egoistic mutant flourishes at the expense of the altruists. If the benefit is translatable into greater fertility (as it must be to make biological sense) then, as the generations pass, the descendants of the egoist will replace those of the altruists. Perhaps not completely – those familiar with genetics will think of the phenomenon of "balanced polymorphism" – but the egoists will become the "type" as altruists diminish in relative frequency to become no more than rare variants in the population.

Mutation and selection, inescapable and ubiquitous, make pure altruism unstable. Our attention must, then, be turned to impure altruism, to the other-serving actions of an individual that in some way serve himself as well.

The best known other-serving action is parental care. That this is not pure altruism becomes obvious the moment we shift our focus from the individual to his or her genes. By caring for his young the parent increases the probability that his genes will survive to remote generations. This care may result in some loss to the parent, in some instances to the greatest loss imaginable, the loss of the parent's life. There is a species of cricket in which the mother permits her numerous brood of offspring to eat her up, thus getting a good start in life.[2] At the genetic level, however, it is not at all altruistic. The mother cricket does not permit any young cricket that happens to be around to eat her. Those who eat her are her own children, and carry her genes. The mother's self-sacrifice is not "for the good of the species"; rather, it serves the good of her germ-line. The genes that cause her to behave in this way are, in a genetic sense, behaving selfishly. This is the insight that led Richard Dawkins to entitle his book, *The Selfish Gene*.[3] Some people regard the term "selfish genes" as a perversion of language, but significant new insights often put old language on the stretch.

It is an irony of history that the term "altruism" was no sooner coined than the pure form of it was shown to be non-existent. Just six years later, in 1859 to be exact, Charles Darwin, discussing the possibility of one species acting altruistically toward another, wrote in his *Origin of Species*:

> *Natural selection cannot possible produce any modification in a species exclusively for the good of another species. . . . If it could be proved that any part of the structure of any one species had been formed for the exclusive good of another species, it would annihilate my theory, for such could not have been produced through natural selection.*[4]

PHOTO: United Nations/Gamma

Persistent pure altruism is impossible not only between species, but also within a species, as the earlier thought-experiment showed. Darwin realized this, as is evident in scores of passages in both the *Origin* and *The Descent of Man*, though he nowhere expressed the point in a brief and quotable way. Nevertheless it is not too much to say that the entire literature of the currently fashionable topic, "sociobiology," is an extended gloss on Darwin.

ALTRUISM AND THE ENVIRONMENT

Sociobiology has been one of the stimulants to a revival of interests in altruism; another has been the concern for the "environment" which has burgeoned in the past two decades. The exact denotation of the word "environment" is often far from clear but discussions of environmental problems seldom continue for long without demands that individuals set aside their selfish desires in favor of the needs of their contemporaries, posterity, or even of an ill-defined "environment."[5]

In general, environmental goods and the costs of environmental abuses are shared by many people, usually without consent. Environment is a common good (or a common bad). Actions, however, have to be carried out by individuals. Proposing that the individual work for the common good raises old questions about the care and nature of altruism. Must the individual sometimes act against his own interests to achieve the common good? Or will self-serving actions suffice?

In the economic context Adam Smith is widely (though not correctly) thought to have answered *Yes* to the last question.[6] His model of the "invisible hand" works well enough (in the absence of monopoly and collusion) to insure that enterprisers sell at the lowest price: seeking their own interest they unintentionally serve the public interest. But the invisible hand fails to prevent ruinous soil erosion when each farmer seeks only his own (short-term) interest, as the history of America's "Dust Bowl" has shown.[7] People often must act in concert (generally, though not necessarily, through government) to bias the free enterprise system so that self-interest becomes congruent with public interest. In general, environmental problems that have not yet been solved are ones that still await the political and social engineering needed to bring about such congruence. Willing assent to engineered changes in the political system requires that many egos be concerned with something other than their *immediate* self-interest. Putting the matter in personal terms, my long-term interest is an interest in my future self, a self who may never be because of intervening death. This future self is a sort of "other"; certainly its interests can conflict with those of my present self. Posterity is another sort of "other"; it too is often served only by some sacrifice of present interests. Concern for the environment cannot be separated from the problems of altruism.

At the most superficial level of analysis, the best of all conceivable worlds for a conscienceless egoist is one in which his egoistic impulses are allowed full reign while his associates are urged to behave altruistically. Unfortunately for the egoist's dreams, symmetry causes other actors to hold the same view. The conflict threatens to produce a stalemate in a world made up of egoists only. But *our* world is not in stalemate so it must not be made up of wholly egoistical individuals. There is at least the appearance of a great deal of altruistic activity, and the appearance needs to be accounted for.

ALTRUISM AND DISCRIMINATION

We easily make sense of other-serving actions once we abandon the search for pure altruism and look for modified or limited altruisms. A significant advance was made when the term "kin altruism" was coined as a name for gene-selfish-individually-altruistic actions, like that of the mother cricket.[8] The central characteristic of all forms of altruism is this: *discrimination is a necessary part of a persisting altruism.* A few examples, from among thousands that could be cited, will illustrate this point.

A bird does not take care of eggs until it has laid its own. then it does not care for just any eggs but only for those in its own nest; and the nest has to be in the right place. If an experimenter moves the nest a few feet, even though the bird sees the action, it will not sit on its own eggs in its own nest once the total *Gestalt* fails to match that demanded by the genetic program in its brain. Caring *and* discrimination are both genetically programmed.

In some species the male helps in the feeding of the young. If the father is killed the mother soon takes on a new consort. The new male ignores nestlings until (1) the offspring of his "wife" have grown up and left the nest, and (2) he has had a chance to mate with the female, who then produces a new family. In human terms, the bird doesn't give a hoot for "his" step-children.[9] Quite a few words are required to state the necessary discriminating characteristics, and our description is probably never complete, but heredity manages to "write" all these discriminations into the genetic code.

Language is treacherous. We are tempted to say that a bird is programmed to take care of "his" or "her" offspring. This would be strictly true only if the individual bird were miraculously capable of recognizing his or her offspring, an ability that technological man, with all his scientific instruments, still cannot do with certainty. What a parent recognizes is a complex sequence of phenomena that identifies, with nothing more than a high degree of probability, offspring that are probably his own.

That this is the correct interpretation of the facts is shown by the success of the cuckoo-bird in exploiting the discrimination system of another species. A cuckoo lays its egg in a nest of the "host species," thus taking advantage of the fact that the host bird does not really recognize its own eggs, reacting merely to eggs of an appropriate size and appearance found in the proper place. When the young cuckoo hatches it proves to be far from altruistic: it grows faster than the young of the host and soon pushes the host nestlings out of the next, thus securing all the parental care for itself.

Is altruism inherited? Yes, but it must be analytically decomposed into inherited helping behavior and inherited ability to discriminate. Among non-human animals with limited intellect, analyzing altruism into these two components may seem rather academic, but for the human species this analysis is of the utmost importance.

Culture, a by-product of inherited intelligence, can modify the inherited rules of discrimination almost without limit. Culture is extra-genetic: it is transmitted from generation to generation by tradition (principally through words). Culture mutates in ways that are quite different from gene mutation.

A complete catalog of all the ways in which human beings have coupled discrimination with caring would be unwieldy. Nevertheless we need some sort of map through the jungle. I present here a grouping of discriminating altruisms that includes the most important altruisms of our time (see Fig. 1).

The various behaviors are arranged in the order of their inclusiveness. At the bottom of the list is *egoism* of the purest sort, a non-altruistic behavior in which the individual literally cares only for himself. In its pure form egoism is non-existence. We are social animals of necessity. (If nothing else, parents must take care of children.) But the concept of pure egoism is a useful base for the assemblage of altruisms.

Immediately above egoism comes *individualism*. It may not be immediately evident that individualism differs from egoism, but individualism can be viewed as the most limited form of altruism. The individualistically oriented person *does* care for others, but mostly on a one-to-one basis. "Love thy neighbor as thyself" is the ideal of an individualistic altruist.

Universalism (Promiscuous altruism)
Patriotism
Tribalism
Cronyism (Discriminating altruisms)
Familialism
Individualism
Egoism

Fig. 1. Egoism and the varieties of altruism, arranged by size of group. In a rough way, the historical sequence is as given, with the older categories toward the bottom of the list.

Dealing with his neighbors one-by-one, the individualist could theoretically include the entire world within the circle of his discrimination. In practice, the circle is far smaller, leading to the rhetoric of individual "rights" which often work against the common good.[10] It takes cooperative action under a majority rule to provide for a national defense force, municipal sewers and mandatory smog control devices. "Libertarians," the most extreme doctrinaire individualists of our time, have difficulty accepting the necessity of any altruism more inclusive than individualism.

Familialism is the term for the altruistic care that family members take of one another. Beyond parental care, familialism is not nearly as important in contemporary America as it is in other parts of the world. In India, for instance, the family is the greatest reality of social existence. In Indian competition strong family ties and obligations are a necessity for individual survival. Indians regard nepotism as perfectly normal and ethical behavior. They are not alone in this. Familialism is powerful in every poverty-stricken, socially chaotic society. So far have Americans departed from time-hallowed familial discrimination that we have even passed laws against nepotism. When the Italian-derived Mafia practices a strong form of extended familialism on American soil we regard this as distinctly unfair, even when their activities are perfectly legal *per se*.

Cronyism is a form of altruism in which discrimination is made on the basis of long association, regardless of genetic relationship. The word "crony" is derived from a Greek word for long-lasting. Cronyism is an adaptive response to the anxiety-creating question, "How can I trust the *other*?" The extensive literature on "The Prisoner's Dilemma" attests to the importance of this question.[11] Because of the "egocentric predicament" *I* can never really know what goes on in the mind of the *other*.[12] Siblings may grow up blessedly untroubled by mutual doubt, but strangers do not enjoy this luxury. Cooperative work, particularly when combined with suffering, creates trust. This is why battle-tested military squads are many times more valuable than green squads. Cronyism then approaches brotherhood; the discriminative delight of it is well expressed by Shakespeare's King Harry:

> *We few, we happy few, we band of brothers;*
> *For he that sheds his blood with me*
> *Shall be my brother.*[13]

The perils of social and commercial life are different from those of the battlefield but they are just as real: they too can nurture cronyism. Not only must cronies trust each other, but in the disorderly maelstrom of civic competition cronies often stand together against the rest of society. The mutual loyalty of cronies in government bureaus and business enterprises often neutralizes the public-spirited actions of "whistle-blowers" who seek to serve the common good by informing against work-associates whose actions violate public laws. Expecting praise, whistle-blowers are more often rewarded with abuse and exile.[14]

The crony-bias of adults has important roots in early childhood. We praise "good citizenship" to our children and proclaim the merits of serving the public; but at the same time we teach the young to detest, loathe, despise, abhor and condemn the "snitch," the informer, the tattle-tale, the squealer and the "stool-pigeon." Where in all these condemnatory words is there a hint of the public interest? The two kinds of messages we give our children are incompatible. Faced with dissonant pressures in adult life the individual, more often than not, favors his cronies against the common good. Both biology and education are responsible for the resulting miscarriage of justice.

The way of the transgressor against cronyism is hard, as the following example shows.[15] Beginning in 1966 officer Frank Serpico tried to reform his corrupt branch of the New York police department from within. After four years of failure he took his story to the *New York Times*. Publication led to an official investigation and the resignation of many high-ranking officers. Serpico, regarded as a traitor by his fellow-officers, was shot in the face and almost killed in a police raid. The circumstances of this event were highly suspicious. In 1972 Serpico went into voluntary exile in Europe and did not return until 1980.

Economic determinists might regard the loyalty of cronies in business as springing solely from mercenary motives. Economic self-interest certainly enters into the conscious or unconscious calculations of cronies but it surely is not the sole motive. When the member of a business team voluntarily leaves to join another firm the severance is usually final. If he becomes disenchanted with his new position he knows, or is soon told, that he cannot resume his old position. Such is the case at least nine times out of ten. His defection is viewed as a rejection of shared values; his formers feel themselves spurned by his departure and find it hard to regenerate their old trust in him. The erstwhile crony is perceived as an apostate: the benefits that might come from re-association seldom seem enough to take the risk. We will accept great objective losses before we will condone or forget apostasy. The spirit of revenge is sure evidence that human beings are far from being pure, or purely rational, egoists.

Tribalism is altruism operating within a tribe, a unit that defies easy definition. Tribal members need not be close kin, nor need they all know each other. They are usually of the same race, but need not be. They share common beliefs, particularly of the sort we call religious. They have the same enemies and react to the same threats. Almost always they speak the same language. They may share geographic territory with other tribes, but if they do they do so in a segregated way. Tribalism is the great reality that has interfered with the development of modern nations in Africa. Africans themselves are acutely aware of this, as one quickly learns by reading their newspapers.

Until recently tribalism has been a very minor kind of altruism in America, but some observers now see the rise of ethnicity and the insistent preservation of multilingualism as signs that America, is moving into a tribalistic phase. The bloody conflict in Northern Ireland and the threat of national fission in Belgium are also viewed as tribalism on the rise. It should be noted that since the founding of the United Nations in 1945 there has been much fissioning of nations and no fusion. It would be naive to suppose that the days of tribalism are over.

Patriotism is nation-wide altruism. I prefer this term to "nationalism," the connotations of which are now so unfavorable as to discourage objective inquire. Even "patriotism" is in some bad odor. Later I shall argue that patriotism can be a virtue. For the present, let us pass to the last and most inclusive altruism, namely universalism.

UNIVERSALISM, THE GRAND ILLUSION

Universalism is altruism practiced *without discrimination* of kinship, acquaintanceship, shared values, or propinquity in time or space. It is perhaps shocking, but entirely accurate to call it *promiscuous altruism*. Its goal was aptly expressed by an unknown poet soon after the end of World War I:

> *Let us no more be true to boasted race or clan,*
> *But to our highest dream, the brotherhood of man.*[16]

The roots of universalism are to be found in the writings of philosophers and religious leaders thousands of years ago, but the promiscuous ideal was given a great boost by the *generalized* idea of evolution in the nineteenth century. W.E.H. Lecky (1838-1903), in *The History of European Morals*, wrote: "At one time the benevolent affections embrace merely the family, soon the circle expanding includes first a class, then a nation, then a coalition of nations, then all humanity . . ." From this passage the contemporary philosopher Peter Singer derived the title of his book, *The Expanding Circle*.[17] Singer believes, of course, that total universalism is not only praiseworthy but possible – perhaps even inevitable.

Universalism is commonly coupled with the political ideal of a world state. The fatal weakness of this dream was pointed out by Bertrand Russell: "A world state, if it were firmly established, would have no enemies to fear, and would therefore be in danger of breaking down through lack of cohesive force."[18] By his phrase "if it were firmly established" Russell indicates that he has carried out a thought-experiment of the sort described earlier in demonstrating that a universally altruistic species could not persist. Russell "pulls his punches" however in saying that a world state would merely be "in danger of breaking down." In fact, it would be certain to break down.

To people who accept the idea of biological evolution "from amoeba to man," the vision of social evolution "from egoism to universalism" may seem plausible. In fact, however, *the last step is impossible*. The forces that bring the earlier stages into being are impotent to bring about the last step. Let us see why.

In imagination, picture a world in which social evolution has gone no farther than egoism or individualism. When familialism appears on the scene, what accounts for its persistence? It must be that the costs of the sacrifices individuals make for their relatives are more than paid for by the gains realized through family solidarity. In the aggregate, individuals who practice familialism have a competitive advantage over those who do not. That is why the step from individualism to familialism is made.

The pattern of the argument just given is characteristically biological, but it is essential to realize that it does not depend on the genetic inheritance of differences in behavior. It assumes no other inheritance than that of the impulse to help and the ability to discriminate. Both impulses can be presumed to be nearly universal in the species. That inherited *differences* are not required by the argument is shown by the following thought-experiment. Assume a random exchange of children resulting in all children being raised by foster parents. Culture alone can be assumed to dictate who does, and who does not, behave familialistically. If familialism is competitively advantageous over the lesser form of altruism (individualism), then familialism will persist. Since biology need not be invoked to account for this cultural step there is no reason for anti-hereditarians to take umbrage at the thought that familialism confers a selective advantage to its practitioners ("selective" being understood in the broadest sense).

Note also that a "higher" grade of altruism does not necessarily extinguish the grades below it. The word "environment" is a singular noun, but the actual social environment in which people have their being is a mosaic of many micro-environments, complicated beyond our ability to capture it in words. In some "spots" individualism will confer an advantage over familialism, in others the reverse is true. If this were not so social life would not exhibit the mosaic of behaviors that it does.

The argument that accounts for the step to familialism serves equally well for each succeeding step — except the last. Why the difference? Because the One World created by universalism has — by definition — no competitive base to support it. Familialism is supported by the competition of families with each other (which favors those with the greater family loyalty) and by competition of families with simple individualists. Similarly tribalism is supported by competition between tribes, and by competition of tribal individuals with individuals who give their loyalty only to smaller, less powerful groups. But those who speak for One World speak against discrimination and for promiscuity: "Let us no more be true to boasted race or clan." What in the world could select for global promiscuity? Only – as science fiction writers have often pointed out – the enmity (competition) of people from Mars, from other worlds. And if the unifying factor of an external threat were to come into being, it is highly probable that the idealists who now speak out for One World would then agitate for One Universe. Evidently what these idealists dislike is discrimination of *any* sort. Unfortunately for their dreams, the promiscuity they hunger for cannot survive in competition with discrimination.

Universalism is truly the Grand Illusion of many in the community of "intellectuals" in our day. How did it get established? This is a fascinating subject for scholarly research. Let me contribute a few pages to the monumental work that needs to be written. One of the most significant short documents is a famous passage from John Donne, from the Devotion that Ernest Hemingway drew on for the title of his novel, *For Whom The Bell Tolls:*

> *No man is an island, entire of itself; every man is a piece of the continent, a part of the main. If a clod be washed away by the sea, Europe is the less, as well as if a promontory were, as well as if a manor of thy friends or of thine own were. Any man's death diminishes me, because I am involved in mankind; and therefore never send to know for whom the bell tolls – it tolls for thee.*[19]

This is beautiful rhetoric and clearly the work of an "intellectual," as we now use that term. But what is an "intellectual"? Alas, it is all too often a person skilled in words but deficient in the imagination required to see the reality behind verbal counters. Consider carefully the images Donne's writing calls forth in the attentive reader. Imagine a promontory, say a cliff at the edge of the sea. If the pounding waves wash away a whole cliff is the loss no greater than if a mere clod were to be washed away? Clod and cliff are equal? And is the loss *to you* the same in these four cases: your house is destroyed – a clod is destroyed? No man of common sense asserts such absurdities.

Donne's prose is a paeon to promiscuity; on this foundation is the dream of universalism built. Denied are all distinctions between large and small, near and far, mine and thine, friend and foe. Yet we must not forget that for three billion years, biological evolution has been powered by discrimination. Even mere survival in the absence of evolutionary change depends on discrimination. If universalists now have their way, discrimination will be abandoned. Even the most modest impulse toward conservatism should cause us to question the wisdom of abandoning a principle that has worked so well for billions of years. It is a tragic irony that discrimination has produced a species (*Homo sapiens*) that now proposes to abandon the principle responsible for its rise to greatness.

We can understand how this has come about if we divide the proficiencies that education produces into three categories: literacy, numeracy and ecolacy.[20] Extending the dictionary meaning somewhat we may say that literacy is the ability to deal with words, whether written or spoken. John Donne was supremely literate: his evocation of man as a piece of the continent "mankind" at first compels our assent to the proposition that each person must be concerned with the welfare of every other person. In weaving his dialectical web the skilled but purely literate man constantly asks himself, *"What is the appropriate word?"*

The numerate man asks another sort of question: *"How much? How many?"* Numbers make a difference. If there were only one hungry human being in the world, who would doubt that we should feed him? But what if the number of malnourished people is 800 million (as it probably is)? And when the number grows to two thousand million, what then? Is it a matter of indifference whether I give a bushel of wheat to my literal neighbor, or to an equally hungry man twelve thousand miles away? (Remember, energy must be used to transport the wheat, energy which cannot then be used to drive a tractor to grow more wheat next year.) Quantities matter, distances matter, numbers matter.

The person whose education encompasses ecolacy is supremely sensitive to time and to the changes that come with time and repetition. The key question of the ecolate person is this: *"And then what?"*

"Ecolacy," derived from the word ecology, tries to take account of the total system in which reactions take place, including such phenomena as synergy, positive and negative feedback, thresholds, selection and boomerang effects. Do pests threaten our crops? Then, says the non-ecolate person, let us generously douse them with "pesticides." (Note the appropriateness of the word.) But ecolacy points out the error: pesticides select for pesticide-resistant pests. Such selection can ultimately defeat our intent and make the situation worse off than before. . . . Is there a housing shortage in our city? Then let us build more houses — surely this will cure the shortage? Not so, says ecolate man. The city is part of a larger system: building more houses will attract more house-dwellers to the city, leaving the housing situation as bad as ever, and the traffic situation worse.

It becomes ever more apparent that the burning questions of our time need to be subjected to the discipline of the ecolate question, "And then what?" Unfortunately, this question is seen as threatening by many vested interests, none more than the philosophers who habitually deal with ethics in a purely literate way. Ethicists of the deontological persuasion attempt the impossible when they try to solve ethical problems with such dull tools as sin, duty, right and obligation – all words blind to number and time-related processes. Consequentialist ethicists, by contrast, are both ecolate and numerate in their approach, insisting that numbers, time and consequences matter.[21]

THE UNEASY COEXISTENCE OF ALTRUISMS

The plurality of altruisms breeds dilemmas. The character of a culture is revealed in the traditional ways it employs to resolve these dilemmas. No characterization of our culture can be complete without some discussion of a famous statement by the novelist E.M. Forster:

> *I hate the idea of causes, and if I had to choose between betraying my country and betraying my friend, I hope I should have the guts to betray my country. Such a choice may scandalise the modern reader, and he may stretch out his patriotic hand to the telephone at once and ring for the police. It would not have shocked Dante, though. Dante places Brutus and Cassius in the lowest circle of hell because they had chosen to betray their friend Julius Caesar rather than their country Rome. . . . Love and loyalty can run counter to the claims of the state. When they do – down with the state, say I which means that the state would down me.*[22]

Forster wrote this in 1939, just before the beginning of World War II. By this time many stories coming out of Nazi Germany told how patriotic Hitler Youth often informed on their own parents when the latter were heard to make statements about Der Fuehrer that were less than enthusiastic. Patriotism was given precedence over familialism. The world was shocked.

As Forster's final sentence implies, patriotism is theoretically capable of overwhelming altruisms of lesser scope. Why does it not *always* do so? Forster said it was because "loyalty can run counter to the claims of the state." The matter can be put more strongly and in quasi-numerate terms: the power of loyalty is *inversely* proportional to the size of the altruistic group. In contrast, political power to control and repress is *directly* proportional to the size of the group. The opposition of the two powers is indicated in Figure 2.

The ineradicable opposition of small group loyalty to the sheer political power of large numbers confutes the supposed drive toward universalism. Because of the egocentric predicament the inference of sincerity in the "other" is always risky, and the greater the number of "others" in a group the greater the risk. The power of loyalty is deeply rooted in innate biological responses to propinquity and repeated association. The power of loyalty to the few constantly erodes the political power of the many. Patriotism depends more on intellectual arguments than does cronyism: this is a key weakness of patriotism. This inherent weakness helps explain the adaptive significance of the theocratic state which proclaims the "divine right of kings." Whenever the support of a state can be made a divine imperative, patriotic loyalty is removed from the realm of rational doubt and shielded from the corrosion of cronyism.

	Universalism	(Promiscuous altruism)	
	Patriotism		
Increasing	Tribalism		Increasing
Political	Cronyism	(Discriminating altruisms)	Loyalty
Power	Familialism		Power
	Individualism		
	Egoism		

Fig. 2. The conflict of powers that works against stabilization at any single level of altruism.

Do the opposing forces create an intermediate point of stability? This seems unlikely. The life histories of individuals vary immensely; the relative valence of political power and loyalty power in the character of each individual is determined by his particular experiences. A crude statistical average might be made for each culture, but there is no reason to think the average would be stable. History forever foils the social systems of the world. Compare the England of Rudyard Kipling with England in the 1930s with its pacifistic "Oxford Oath" taken by millions of young men. The Boer War and World War I moved the statistical balance point of the discriminations "downward" (on the list in Figure 2 – no ethical interpretation is implied). Then when Germany invaded Poland in September of 1939 the Oxford Oath was abruptly jettisoned and the balance point moved decisively "upward" toward patriotism. It has since fallen in England. In America it has fallen even more, as a result of the Vietnam war. The manifest dangers of nuclear war argue (to some) for a permanent abandonment of patriotism, but the argument is valid only if there are no reasons *other than war* for supporting discrimination at the national level. We will return to this point later.

THE MISSING MIDDLE TERM

"Liberalism" is an ill-defined term of constantly changing meaning, yet (whatever its meaning) it is not far off the mark to say that liberalism enjoyed more praise than power in the nineteenth century, whereas now it enjoys more power than praise. Hell, as someone said, is when you get what you want. With power, self-doubts have come to the liberals. The fashionable journals of the literate world are now pulsating with liberal threnodies.

The political philosopher Michael Novak has put his finger on a key weakness of what is, in our time, called liberalism:

> *The liberal personality tends to be atomic, rootless, mobile, and to imagine itself as "enlightened" in some superior and especially valid way. Ironically, its exaggerated individualism leads instantly to an exaggerated sense of universal community. The middle term between these two extremes, the term pointing to the finite human communities in which individuals live and have their being, is precisely the term that the liberal personality disvalues.*[23]

That liberals should regard themselves as elite — literally "chosen" — means nothing more than that they are human. They enjoy an *esprit de corps,* a feeling which those outside a chosen circle identify as ethnocentrism (a sin, be it noted, especially deprecated by contemporary liberals). What needs explaining is the apparent paradox (irony, Novak calls it) of combining individualism and universalism in the liberal personality, with no "middle term."

In the assemblage presented in Figure 2, Novak's "middle term" is decomposed into four different altruisms. Of these, the most conspicuously lacking among contemporary liberals is patriotism. Forster's condemnation of this form of altruism could easily be matched by hundreds of other statements coming from the liberal, "intellectual," literate community. Patriotism has had a bad press ever since Dr. Johnson's offhand remark, "Patriotism is the last refuge of a scoundrel."[24]

Never has the defense of individual "rights" been as strong as it is in our time. Why, then (to paraphrase Novak), does exaggerated individualism lead to exaggerated universalism? To a biologist this puzzle presents little difficulty. Among altruisms, individualism is clearly a borderline case; psychologically it is close to naked egoism. *Homo sapiens* is a social animal: his social appetite is not completely satisfied by an altruism that goes no farther than the *I-Thou* relationship of Martin Buber.[25] Our groupish hungers are seldom completely satisfied by purely dyadic relationships. A significant fraction — perhaps even a large fraction — of humankind craves identification with groups larger than *I and Thou.*

Radical individualism is often linked to hedonism. One sees this clearly in the multitude of magazines in the *Playboy* mode. A practicing playboy is not a complete egoist because "it takes two to tango," but his individualism is of a low order, for the *other* is little more than a sex object. In the past, women (more than men) may have been the guardians of community values; now there is a *Playgirl* magazine that seeks to erase the difference. For Americans, the Declaration of Independence has supplied a banner for hedonism: "the pursuit of happiness."

Hedonists of both sexes should be informed of what the nineteenth century philosopher Henry Sidgwick called the *Hedonistic Paradox*: those who most actively pursue pleasure as a primary goal are least likely to achieve it. Personal happiness is best gained by indirection, by serving some larger cause. I think this can be taken as an empirical fact. By way of theoretical explanation I would point to two factors.

First, since we are social animals who find pleasure working with others, the horizon of our attention must be broadened beyond the bounds of egoism; perhaps the greater the cause the greater the pleasure in serving it. Secondly, human beings find so much pleasure in overcoming difficulties that they even seek out difficulties to overcome. We climb mountains that stand not in our way – and thus discover new ways to happiness. Behavior that to a simple rationalist might seem perverse plainly has contributed to the success and progress of the human species. Progress has selected for temperaments that find the simple hedonism of unalloyed individualism too low a peak for complete satisfaction. Not all human beings transcend the demands of simple hedonism, but enough do to affect the course of history. To forego short-term hedonistic gain for a dream that may – only *may* – be realized in the future is to fall into a behavioral pattern that supports altruism.

The dreams of today's more far-seeing individuals are most commonly universalist dreams: One World, the Brotherhood of Man and the like. Though universalists disparage the moral value of lesser groups, in furthering their cause they necessarily rely on cronyism. Ironically, cocktail parties to which liberals alone are invited are a great place to denounce elitism, the enemy of promiscuity. Thus is the cause of promiscuity advanced by discrimination.

All causes succeed through close-knit, small groups. The effectiveness of a great army, serving patriotic ends, is determined by the cronyism of multitudinous small squads (a fact long recognized by the military). Similarly, the effectiveness of liberals in pursuing universalist ends is determined by the cronyism developed in small groups. The grass roots of patriotism and universalism are the same, only the ends differ. Why has patriotism been rejected by contemporary liberals? It is to this that we now turn our attention.

SHARING THE ENVIRONMENT

The universe may or may not be finite, but prudence demands that we assume that the portion *practically* available to humankind is finite. Technology effectively expands this portion somewhat, but at a rate that is less than the expansion of our expressed demands: hence the unending complaints of "scarcity." The analytical model for productive economic thinking must be that of a "closed system," a system in which input matches output (diminished somewhat by entropic loss). The enduring task of political economy is the allocation of scarce resources.

No sizeable, prosperous society has been able to persist for long under a rule of equal distribution of income, wealth or privilege. This empirical fact has not interfered with the persistence of the dream of distributing goods by the rule, "to each according to his needs" – to use Marx's language for an ideal furnished him by the religion he despised.

Empiricism is not enough: before we can assent to an apparent impossibility we must "understand" it, that is we must find the theoretical "impotence principle" that sets the limits.[26] Why won't a Marxian distribution work? To answer this we must ask, distribution of what? It makes a difference.

The "what's" of the world come in three varieties: matter, energy and information. Every redistribution of matter and energy is in accordance with zero-sum principles: the gain to A is exactly matched by the loss to B. Equations must balance: the mass (or quantity of energy) on the left side must match that on the right. Matter is conserved. Energy is conserved.[27] Matter and energy obey "conservation laws."

However great our social impulses, evolution has selected for an irreducible minimum of egoism. Any proposal to transfer the goods of matter and energy from B to A is likely to be resisted by B.[28] Overpowering such resistance uses up "energy," either in the physicist's sense or in some other significant sense. It is highly doubtful that there ever was any "initial state" of equidistribution of human wealth or social power. Equidistribution, if at all possible, can be achieved only by some impoverishment of the group as a whole – in the case of violent revolution, by massive impoverishment (and an invariable failure to achieve the goal of the instigators). Violence, which accelerates the drive toward entropy, creates a negative-sum game. This is the consideration that moderates the enthusiasm of the prudent man for "distributive justice." Territorial behavior in other animals and property rights among human beings often serve the same cause – the cause of peace.

There are three basic politico-economic systems: privatism, socialism and commonism.[29] Privatism (under various names – "private enterprise," "capitalism," "free enterprise," etc.) never takes equidistribution as a goal, though apologists often assert that a "trickle-down effect" slowly works toward that end. Socialism and commonism, however, seem congenitally committed to the ideal of equidistribution. Under socialism, the major part of the community's wealth is kept as common property which is managed (supposedly) for the good of all by managers appointed more or less directly by the community. This property may be spoken of as a "managed commons."

Under commonism, however, the commons is unmanaged, being left available to all under the Marxist rule, "to each according to his needs." Under conditions of abundance, commonism may work very well. The hunting grounds of the pioneer days of America were a commons that worked. An unmanaged commons has the advantage that the cost of management is zero. But when people become crowded and resources scarce an unmanaged commons does not work well because each individual is the judge of his own needs. With scarcity, commonism favors egoism over altruism. The would-be altruist, if he is to survive under scarcity, must become as egoistic as the worst. In the name of freedom and distributive justice, an unmanaged commons breeds harsh egoism, inequality and injustice. So long as such a system endures men of good will are powerless to change the results: such is the "tragedy of the commons."

The commons that led the obscure English mathematician W.F. Lloyd to deduce its analytical properties a hundred and fifty years ago is now not very important.[30] This was the commons of English pasture land. But the commons of oceanic fisheries and the sea bed (from which valuable minerals can be extracted) still exist and promise to create international trouble in the future. So too does the commons of the atmosphere which serves as a sink for the "bads" of volatile pollutants.

Without being aware enough or honest enough to use the proper label we constantly create new commons. Insurance, which begins as a wager, tends towards a commons as the fraction of people insured approaches unity. Those who are insured pressure the system to make premiums equal while wanting payouts to be made according to unmonitored needs. To keep the costs of automobile accident insurance and fire insurance from ruinous escalation there must be constant monitoring by managers alert to arson and fraudulent repair claims.

Universalism is the ideal of One World in which clod equals cliff; the "rights" of all are equal, whether friend or foe, native or foreigner, relative or stranger. A universalist is, whether he acknowledges it or not, a follower of Marx and a promoter of the tragedy of the commons. How, then, are we to account for Novak's observation that the liberals of our time have hybridized the altruisms at the extremes of the scale, namely individualism and universalism?

The answer is to be found in the peculiar nature of words, the medium of the merely literate intellectuals who are so influential in our time. By words we convey information. Unlike matter and energy, *information is not subject to conservation laws.*

Agent B, in the act of giving information to A, loses nothing. In fact, if

A reworks the information into an improved form and passes it back to B, both gain. Far from being a zero-sum game, information-sharing can be a positive-sum game. When we deal with information there are strong reasons for sharing generously, even for maintaining a commons of information. Science could not have made its rapid progress had information been treated like a property subject to conservation laws.[31] In espousing universalism professional literates are merely generalizing from their profession to the world at large, unaware of the significant difference between information on the one hand and matter and energy on the other.

What, then, accounts for the individualism of this group? This is no secret: "Shakespeare's plays could not have been written by a committee." Creativity at the highest level is inescapably individualistic. There is no "group mind" to carry out the decisive act of creation.

The One World that universalists dream of is clearly a world freed of many of the restraints of lesser political units, a free world ("to each according to his needs"). It is easy for a radical individualist to embrace universalism while rejecting all intermediate altruisms. The strengths of individualism are unaffected by this hybridization of ideals precisely because no real universalist world exists to act as a restraint on the individualist who worships it as an ideal. Were One World to come into being, and were it to acquire the sanctions that all lesser associations have taken unto themselves, the individualist would find himself unhappier than ever. There would then be no larger ideal for him to aspire to.

Universalism is attractive in large part because the ideal is used as a weapon to beat off the restraints necessarily imposed on individuals by family, tribe and nation. In deciding how much support to give individualism we are well advised to examine the "track record" of individualism. Philosophers and historians are pretty well agreed on the meaning of the Greek experience:

> *The greatness of the Greeks in individual achievement was, I think, intimately bound up with their political incompetence, for the strength of individual passion was the source both of individual achievement and of the failure to secure Greek unity. And so Greece fell under the domination, first of Macedonia, and then of Rome. – Bertrand Russell.*[32]
>
> *Individualism in the end destroys the group, but in the interim it stimulates personality, mental exploration, and artistic creation. Greek democracy was corrupt and incompetent, and had to die. – Will Durant.*[33]

It is exciting to live in a world of richly creative people but the individualism that fosters creativity may, *unless it becomes self-conscious,* destroy the foundations of the society that supports it. "Becoming self-conscious" means that "intellectuals" must realize that the One-World commonism they aspire to is only a *natural, though fatal, inference from their craft,* which is the elaboration and distribution of ideas and information. Matter and energy, by contrast, must be distributed with discrimination, not promiscuously, else the tragedy of the commons will be set in train. "Intellectuals" must learn to praise virtues different from the ones that give them their craft-strength. The survival of a civilization in which intellectuals have great social power requires that this power be coupled with a degree of objectivity that is rare among men of all vocations.

Universalism is unattainable, and individualism is not enough – not in a competitive world where a larger group has the edge over smaller ones. The last remark is, of course, to be understood *ceteris paribus;* but the thrust of the argument pushes us toward the conclusion that there will always be an important role for the altruism that is only one step below universalism. That is the altruism we call "patriotism."

Many concerned people today find this conclusion hard to swallow. Patriotism; war, nuclear holocaust; destruction of civilization – this chain of ideas has led many to believe that patriotism must be expunged to save civilization. The establishment of One World is seen as a way to dismantle the armaments of nations. But promiscuous universalism would destroy the world too, though in a different way: in T.S. Eliot's prescient formula, "not with a bang, but a whimper."[34]

The whimper has begun, but so far as I know, only one literary man has noticed the form it is taking: the French writer Raspail in his novel *The Camp of the Saints.*[35] His argument is only implicit (as a good fiction-writer's should be) but it is easy to translate it into explicit stages. The logical steps in the developing disaster are these:

1. By virtue of their craft, opinion-makers worship the ideal of promiscuous sharing: for them, patriotism is unthinkable;
2. "To each according to his needs" means that when immigrants from a poor country knock at the door of a rich country they must be admitted;
3. The process of moving from poor to rich will continue until wealth is equalized everywhere;
4. But since there is no group limitation on individual freedom to breed it is not so much wealth that will be equalized as it is poverty – thus plunging everyone into the Malthusian depths.

Have we no choice other than between the whimper of common pauperization and the bang of thermonuclear destruction? I think we have. I am enough of an optimist to believe that we can create and sustain forms of patriotism based on national pride in the arts of peace — science, music, painting, sports and other arts of living. Excellence in these accomplishments can be the occasion for community pride (hubris, if you will), which has its dangers but without which life is not fully lived. Accompanying all this there must be the patriotic will to protect what has been achieved against demands for a world-wide, promiscuous sharing. A community that renounces war as a means of settling international disputes still cannot survive without that discriminating form of altruism we call patriotism. It must defend the integrity of its borders or succumb into chaos.

IN SUM

The caring impulse, generalized without limit, produces universalism which, though desirable in the realm of information, is destructive when it comes to matter and energy because promiscuous sharing of limited physical resources leads to the tragedy of the commons. Some people have revived the old motto "All men are brothers" with the assertion that the pageant of Darwinian evolution gives it new meaning. Possibly so, but the conclusion that brotherhood requires us to perish in a commons is a *non sequitur*.

If biology is to be consulted for guidance we must take note of this supremely important fact: a species does not survive because its members act "for the good of the species," but because individuals act for the good of themselves, of their germ-lines, or of reciprocity-groups smaller than the total population. The survival of the species is, as it were, an accidental by-product of discriminating altruism. Biologists have known this more or less ever since Darwin, but it has become crystal clear only in the last two decades.

Completely promiscuous altruism in a species that has no important enemies would destroy both the species and its environment. A judicious mixture of discriminating altruisms is required for survival. The universalist's dream embodied in St. Augustine's *City of God* can be realized only in the realm of ideas, which alone can be promiscuously shared with safety.[36] We must be chary of deducing any *material* consequences from the assertion that "All men are brothers." The pleasures of brotherhood are sweet, but only because they involve both caring *and* discrimination, as Proudhon realized a century ago: "If everyone is my brother, I have no brothers."[37]

Brotherhood requires otherhood. Civilization has been built upon, and can only survive with, a changeable mixture of discriminating altruisms.

NOTES

[1]Benjamin Lee Whorf. *Language, Thought, and Reality.* (New York: Wiley, 1956).

[2]Richard D. Alexander, "The Evolution of Genitalia and Mating Behavior in Crickets (Gryllidae) and other Orthoptera," *Miscellaneous Publications of the Museum of Zoology, University of Michigan* 133 (1967): 1-62.

[3]Richard Dawkins, *The Selfish Gene* (New York: Oxford University Press, 1976).

[4]Charles Darwin, *The Origin of Species*, 6th ed. (New York: Macmillan, 1927). pp. 197-198.

[5]Christopher Stone, *Should Trees Have Standing?* (Los Altos, Calif.: William Kaufmann, 1974).

[6]Adam Smith, *The Wealth of Nations* (New York: Modern Library, 1937). p. 423.

[7]Paul B. Sears, *Deserts on the March* (Norman: University of Oklahoma Press, 1980).

[8]The first step in the development of the idea of altruism-as-discrimination was made by Robert Trivers in his article, "The Evolution of Reciprocal Altruism," *Quarterly Review of Biology* 46 (1971): 35-37. Early critics protested that behavior that followed the rule of "You scratch my back and I'll scratch yours" was not altruism at all – which indeed it was not, in the pure sense. Trivers freed our minds from the shackles of the idealist's unreciprocated altruism, thus turning our attention to discriminating altruisms, the essential elements of social existence.

[9]Harry W. Power, "Mountain Bluebirds: Experimental Evidence against Altruism," *Science* 189 (1975): 142-143.

[10]Garrett Hardin, "Limited World, Limited Rights," *Society* 17 (1980): 5-8.

[11]Anatol Rapoport and A.M. Chammah, *The Prisoner's Dilemma* (Ann Arbor: University of Michigan Press, 1965).

[12]I have been told that Ralph Barton Perry (1876-1957) coined the term "egocentric predicament," but I have not verified this. The term is seldom used, which suggests that the underlying phenomenon is under something of a taboo. Social intercourse is facilitated by a belief in the sincerity of the "other," which is an unknowable. Most traditional ethics is concerned with intentions, which are also unknowable. The law wisely is built on actions, but it frequently lapses into inferring intentions, as in the case of "fraud." In our desire to shield our minds from the corrosion of doubt we usually suppress the sure knowledge that we can never know what goes on in the mind of the "other." Social life is permeated with this suppression.

[13]William Shakespeare, *King Henry V*, Act IV, Scene 3, line 60 (1599).

[14]Many disturbing examples of the treatment of whistle-blowers are to be found in Alan F. Westin, ed., *Whistle-Blowing: Loyalty and Dissent in the Corporation* (New York: McGraw-Hill, 1981). For the particular story of a high-level dissident in the General Motors Corporation, John Z. DeLorean, see J. Patrick Wright, *On a Clear Day You Can See General Motors* (New York: Avon, 1979).

[15]Mitchell Satchell, "Frank Serpico is Coming Home," *Parade* (12 October 1980). p. 9.

[16]Thomas Curtis Clark, "The New Loyalty," in Thomas Curtis Clark and Esther A. Gillespie, eds., *The New Patriotism* (Indianapolis: Bobbs-Merrill, 1927).

[17]Peter Singer, *The Expanding Circle* (New York: Farrar, Straus & Giroux, 1981). The quotation from Lecky was taken from this source.

[18]Bertrand Russell, *Authority and the Individual* (London: Unwin, 1949). p. 17.

[19]John Donne, "Devotion XVII" (1624), in *The Complete Poetry and Selected Prose of John Donne & The Complete Poetry of William Blake* (New York: Modern Library, 1946). p. 332. I have modernized both spelling and punctuation.

[20]Garrett Hardin, "An Ecolate View of the Human Predicament," in Clair N. McRostie, ed., *Global Resources: Perspectives and Alternatives* (Baltimore: University Park Press, 1980). pp. 49-71.

[21]A good presentation of this view, written before the word "ecolate" was coined, is found in Joseph Fletcher, *Situation Ethics* (Philadelphia: Westminster Press, 1966). For a more recent discussion see Garrett Hardin, *Promethean Ethics* (Seattle: University of Washington Press, 1980).

[22]E.M. Forster, "What I Believe," in *Two Cheers for Democracy* (New York: Harcourt, Brace & World, 1951). p. 68.

[23]Michael Novak, "The Social World of Individuals," *Hastings Center Studies* 2 (1974): 37-44.

[24]Samuel Johnson made this remark in 1775, at the age of 66. But Johnson was not condemning true patriotism. As Boswell said, "Patriotism having become one of our topics, Johnson suddenly uttered in a strong determined tone, an apophthegm at which many will start: 'Patriotism is the last refuge of a scoundrel.' But let it be considered that he did not mean a real and generous love of our country, but that pretended patriotism which so many, in all ages and countries, have made a cloak for self interest." (*Boswell's Life of Johnson*, Everyman's edition, vol. 1, pp. 547-548.) In other words, patriotism is the last of a scoundrel's many refuges, most of which bear the names of virtues.

[25]Martin Buber, *I and Thou* (New York: Scribner's, 1970).

[26]E.T. Whittaker, "Some Disputed Questions in the Philosophy of the Physical Sciences," *Proceedings of the Royal Society of Edinburgh* 61 (1942): 160-175.

[27]The combination of matter and energy into Einstein's Law, $E=mc^2$, need not concern us here.

[28]Garrett Hardin, "Living on a Lifeboat," *BioScience* 24 (1974): 561-568.

[29]Garrett Hardin and John Baden, eds., *Managing the Commons* (San Francisco: W.H. Freeman, 1977). See particularly chapters 1, 2, 3, 7, 9, 11, 19 and 25.

[30]William Forster Lloyd, "On the Checks to Population," (1833) in *Lectures on Population, Value, Poor-Laws and Rent* (New York: Augustus M. Kelley, 1968).

[31]In passing, we note that sometimes there are advantages to treating information as property. Copyright and patent laws do so, and make it possible for the originators of good new ideas to reap profits, thus encouraging others to be inventive. The transferability of these property rights makes it commercially possible for enterprisers to make the investment needed to convert idea into product, a possibility foreclosed to a public patent (which creates a commons). But property rights in information are difficult to police; note, for example, the pirating of computer software and tape recordings.

[32]Bertrand Russell, *Authority and the Individual* (London: Unwin, 1949). p. 27.

[33]Will Durant, *The Life of Greece* (New York: Simon & Schuster, 1939). p. 554.

[34]The last two lines of *The Hollow Men* (1925) are: "This is the way the world ends/Not with a bang but a whimper." T.S. Eliot, *The Complete Poems and Plays* (New York: Harcourt, Brace, 1952). p. 59. Exercising his right to be elliptical and ambiguous, a poet always leaves us wondering whether he is unusually prescient or merely lucky.

[35]J. Raspail, *The Camp of the Saints* (New York: Scribner's, 1975).

[36]Robert Nisbet, *History of the Idea of Progress* (New York: Basic Books, 1980). p. 60.

[37]Alexander Gray, The Socialist Tradition (London: Longmans, Green, 1946). p. 159.

Homage to Heidegger

Paul Shepard

Naturalist Paul Shepard has been a soldier, fruitpicker, editor and sailor. He began writing and teaching as "an ecologist looking at the human species." His books include Man in the Landscape, The Tender Carnivore and the Sacred Game, Thinking Animals, Nature and Madness, *and (with Daniel McKinley)* The Subversive Silence. *Dr. Shepard is an International College tutor and Avery Professor of Human Ecology and Natural Philosophy at Pitzer College and The Claremont Graduate School.*

I learned about the Hun and the Bosche at my father's knee. The "enemy" for us small boys in our play with toy guns was just as often Germans as it was Italians. My uncle had a 24-volume set of rotogravure picture books from World War I that provided abundant visual fantasy to feed the inner eye in our games of shoot and die. By the time I got to high school we were into World War II, and I served the last six months of the war in Europe with an armored division in France and Germany. My battalion "liberated" the Landsberg concentration camp with its three living survivors and the horror of the Holocaust has not diminished in the decades since then. Even trivial experiences in later years – phalanxes of German tourists turning quiet Spanish cafes into loud cells of sentimental boorishness, or peaceful streets suffering the stench-in-the-ear of Volkswagens – seemed to confirm the utter insensitivity of the "kraut" wherever he appeared. In college I studied ecology (a word said to have been coined by the German zoologist Haeckel) and in class we made fun of the Prussian example of bad forestry: regimented plantings wherein the natural woodland community was so decimated that rodents, swarming in the absence of foxes and owls, rose up and destroyed the young trees. So, for half a century the Germans seemed to me a dark and hopeless people, even though there is a family root on my mother's side of Schwarzes and Webers, ancestors of some distinction in Wurttemburg.

The other thread of this short discourse begins with that ecology course just mentioned. However mistaken the style of German forestry was, we knew that ecology was to be applied to the care of the land. Particularly, I studied wildlife management, a program of applied science aligned in the land-grant universities with range management, soil management, forest management, and so on. It seemed to us in those years, while Aldo Leopold was still living, that a great light was dawning. We understood that ecological relationships in nature were almost infinitely complicated, and yet the principle was so simple: the whole of the living community had to be taken into account in the design of the various "managements." If that were done we would not go on making deserts, eroding the soil, and generally draining the earth of its resources. Such was the idea, or rubric, under which all the earth skills would become a single enlightened resource management. In the refined techniques for sustained production the habitat was to be considered. The managers were also to learn that their activities had side effects. We envisioned wildlife management as the prince of this new nobility of professions, since most of the game lived on land used or (hopefully) managed for something else.

Such were the conservation programs at mid-century. It was to be an educated exploitation in which the spoiler and waster would come to see that professionals so trained could enable them to get more from the earth in the long run. As a youth I had been interested in kinglets, fence lizards, and box turtles, and it bothered me a little that these would not benefit much from management, however ecologically inspired it was. I had doubts about the scope of this approach to "wildlife" and wrote an article called "The Dove is Doubtful Game." published by *Nature Magazine* (now defunct), which infuriated one of my management professors. To a colleague he used the word "traitor."

Feeling that something else was needed, I went off to an even more enlightened graduate school in the Ivy League. There we talked about cultural values and planetary matters like collapsing biomes, overpopulation and biogeochemical cycles. Land use technique alone would not do; we needed to know about social proceses, to see how the laws and taxes sometimes made landskinners of the farmers or coal companies. We had to face up to the steely grip of habit and to the cults of use and waste, the status symbol of consumption and the psychology of growth and progress. Having given up chickadees for game animals, and leaving them in turn for more distinctly human topics, I attempted to carve out a niche in this impossibly broad field with a thesis on the history of land aesthetics. What we neded, I supposed, was to balance out the practical use of nature against its intangible values. Even in the materialist West there was a heritage of gardening, travel, and painting. The idea of the landscape fascinated me; it seemed to be the aesthetic equivalent of the ecosystem. If ecology was too esoteric for the public to understand – this was 1952 – perhaps the notion of scenery was a built-in response to the same kind of thing.

It took me more than a decade to work my way through the landscape. I owe my liberation from it to the work of geographer David Lowenthal* and Marshal McLuhan. Their articles convinced me that the world-as-picture was on one hand geared to the superficiality of taste and on the other an outcome of a Renaissance mathematical perspective that tended to separate rather than join the human and the nonhuman. The landscape was an inadequate nexus. It was only a twist in a substratum of ideas of the earth older and more profound than I had thought. Indeed, such ideas depended as much on unconscious perception as on intellectual or artistic formulations. I began to feel that something more biogenic, something common to mankind, which yet might take particular social or aesthetic expression, held the key to an adequate human ecology.

Over the next decade I read anthropology and child psychology. During that time a meeting of anthropologists took place in Chicago that resulted in the publication of *Man the Hunter*. I began to think that the appropriate model for human society in its earth habitat may have existed for several million years. If Claude Levi-Strauss were to be believed, nothing had been gained by the onset of civilization except technical mastery; what had been lost or distorted was a way of experiencing in which nature was an unlimited but essential poetic and intellectual instrument in the achievement of human self-consciousness, not only in evolution but in every generation and every human life. I knew such an idea would be ridiculed as a throwback to nostalgia for the noble savage, but when it was considered in the light of Erik Erickson's concept of individual development as an identity-shaping sequence, I found it irresistible.

The essence of prehistoric human society could be better appreciated when seen against that which replaced it: the peoples of plant and animal domestication, of agriculture, pastoralism, and the growth of tribes, towns and wars of the Near and Middle East; and more than that, their catastrophic destruction of the land coupled with ascetic and abstract

**I hated him for it at the time.*

philosophies of transcendence in a "world" religion. The subject was a bit large for a devotee of box turtles.

Still, there are intimations that might be worth passing along. Ideology and history do not seem very important in this connection – in terms of their content – but ideologizing and historicizing are it its heart. They replaced the mythological foundation with its atochthonous roots. Therefore, the focus is not on good and bad ideas, nor even on the perceptual processes associated with them, but on something unidentified, on a form of experience so fundamental that no idea alone can replace it. The destructive effects of the temporal mode of history are somehow inflicted at an early age on every child in the western world. One wants to call it Hebrew or Greek or Christian but, judging from the wreckage of the land and its life in their own world the Buddhists and Hindus offer no helpful alternative.

The only alternative visible to me now is northern. The poisoned ontology-ontogeny carried into Europe by Christians between the fall of Rome and the Reformation was diluted again and again by the pagans and heathens who were, ostensibly, converted. The *desert mind*, a Platonic, prophetic, self-centering, dualistic, schizoid, eco-alienating way of being, could not have been less like the Germanic, Celtic, and Scandinavian *way* that it eventually quashed. As a Japanese philosopher, Watsuji Tetsuro, has remarked, it is astonishing that a hemisphere of people in the north could believe that their whole existence hinges on things that happened to a small, distant, desert people two millennia ago. But the box-turtle student's reaction is to ask whether they have indeed come to believe it or whether ten millennia of tampering with the way in which we individually perceive ordinary experience has not been managed so as to make it seem so.

In that case, the Protestants might be seen as hard-liners who could see what the northern pagans were doing to Christianity, the compromises which it was making in order to gain a foothold, the whole infiltration into the orthodox church as described by Seznec in his book, *The Survival of the Pagan Gods*. Like the Indians of North America centuries later, who went underground with their religion as the Protestant American government sought systematically to destroy their heritage, the Lapps, Finns, Hungarians, and Germans retained their "superstitions" in private and brought them masked to church. The church prelates, in what they thought was a strategy of assimilation, kept the polytheistic holidays; but their success was their own perversion. It would be interesting to know today whether the good land-use which makes much of Europe so beautiful is due to customs that survive secretly, despite the other-worldly *way* that has dominated the North for so many centuries.

So, now we come back to those hateful Germans with these thoughts. Techniques, however centrally organized or scientifically sophisticated, will not alone preserve the productivity and health of the planet. At a slightly deeper level, neither will alterations in law and government, economic system, political strategy, or media – although they may yield remedial effects. They may save a species today, clean a river, reduce a smog, do some other temporary and piecemeal good. But they cannot even describe, much less solve the problem. Going still below these social realities, we may suppose that our behavior springs from philosophical systems or religious doctrines, even in those of us who cannot articulate and seldom think about them. Such systems infuse the whole of a society with assumptions, interpretations and implicit directions. Yet, for me, such a set of connections and pervasive outflow from reasoned theory do not explain adequately the way in which the youngest and uneducated members of society join so readily in its major thrust and find alternatives so difficult to comprehend. A contemporary philosopher who tells us that we must somehow find a still more basic mode below willed thought is a German, Martin Heidegger.

And what is it about Heidegger that would make an Ozark ecologist forgive his German cousins? The professional philosophers cannot agree among themselves what he says, so I can be rash, make what I can of that part of his work available in English, and take some cues from George Steiner, the pungent stylist and author of a book on Heidegger. Primarily, Heidegger makes constant connections to the biology of man in its broadest sense. Without building "up" from the natural sciences, he seems to move by a strange path and intuition toward certain psycho-ecological realities (in somewhat the same way that José Ortega y Gasset did); that is, by undercutting and rethinking many of the premises and givens of traditional Western thought.

He is obsessed with language and words, how they mean in a sense fundamental to experience and to our attention rather than as the external decoration of thought. His constant metaphors of movement through a terrain of the mind do not come across as arbitrary images of literary humanism but rather as a function or process of intelligence that evokes both a sense of origins in a hunter-forager past and the instrument by which each growing child passes toward self-identity. Yet it is not a self– or even human–centered approach. Steiner remarks, "For Heidegger, on the contrary, the human person and self-consciousness are *not* the center, the assessors of existence. Man is only a privileged listener and respondent to existence. The vital relation to otherness is not, as for Cartesian and positivist rationalism, one of 'grasping' and pragmatic use. It is a relation of audition. We are trying to 'listen to the voice of Being.' It is, or ought to be, a relation of extreme responsibility, custodianship, answerability to and for."

Heidegger insists on our awareness of the mystery of being and the numinous power of the words for being. His challenge to the thought of

Aristotle, Plato, Bacon, Descartes, Hegel, Kant and the other luminaries of the analytic and positivistic "wilful sovereignty" of the ego is not even undertaken in the context of the transcendental Western metaphysics at all, and that, I suppose, is why it is so difficult. He seeks to oppose the "imperialist subjectivity" by seeking the defect in something prior to its logic.

"To rethink in this way," says Steiner, "a man must repudiate not only his metaphysical inheritance and the seductions of 'technicity,' but the egocentric humanism of liberal englightenment and, finally, logic itself... True ontological thought, as Heidegger conceives it, is presubjective, prelogical and, above all, open to Being. It *lets Being be*. In this 'letting be' man does play a very important part, but it is only a part." Try that on your Buckminster Fuller program.

In a world where the technician is no longer a craftsman, Heidegger pays attention to "the existential fabric of everyday experience, the seamless texture of being which metaphysics had idealized or scorned. ...Any artist, any craftsman, any sportsman wielding the instruments of his passion will know exactly what Heidegger means and how often the trained hand 'sees' quicker and more delicately than eye and brain."

Such an idea is for me like a crash of cymbals, triggering images of the marvelous Magdalenian tool kit, fifty or sixty or a hundred kinds of finely made instruments of bone, tooth, antler, stone and probably wood that were swamped in Neolithic times by the coarse implements in the deformed hands of men enslaved to shovels, harnesses, hoes and plows. It helps me to understand what the poet Gary Snyder means by "work," not a drudging, slogging subservience to a hated routine but an encounter with the world in a delicate way central to thought and feeling.

Heidegger's feeling for the sanctity of the earth is not based on a penitential ecology, reparation, pseudo-theology or political radicalism, says Steiner; not an increase in knowledge or mastery, but a "collaborative affinity with creation," which calls even, if you will, for a renewal of ancient deities and the play of the agencies of vital order as they were seen in "pre-Socratic Greece."

For my part, the route Heidegger takes is not just the recovery of a rootedness like that sustaining the minds or bodies that built the archaic Greek temples, but something alluded to and never formulated in Claude Levi-Strauss's *The Savage Mind*, something one may sense in the broken remains (or visible tip) of American Indian reflective thought, or the possibilities suggested in André Leroi Gourhan's *Treasures of Paleolithic Art* — the religious meaning of rock shelter and cave art. It is not recoverable entirely by what is usually meant by "philosophical speculation." Yet it is latent in the way in which speech and sensory experience come into their own in the life of the child. In its fruition, that "something" is not a childlike phenomenon, although it exists potentially in each child.

Somehow there is a wrong-turning in what is done in childhood, an error repeated as we make a world (to paraphrase Winston Churchill) and it then makes us, a failure in nurturing done again by the children as they become parents.

References of a Heideggerian kind to earth, place and blood release in us a deep fear of Teutonic barbarism, the Hun horseman, slaughters of whole generations of Europeans and Mediterraneans by the invading peasants and pastoralists from the North and East. The Christians found it easy to make them the compleat example of terrible heathen. Indeed, there is no denying the historical facts. It has been said so often as to become trite that the Germans manage to embody the worst and the best. But from what position do we view this awful ambiguity? We – and probably the Germans themselves – have re-envisioned that savage aspect through idealized, Mediterranean, desert lenses. It is a bad lens that in the interest of analyzing leaves only a prismatic scattering and shattered subject. A hypocritical lens. Indeed, in that northern ferocity there is a big contribution from the hot mind of Holy Land ideologists who have been killing each other (and exterminating other species) for little enough reason for six thousand years.

Right now I am going bird-listening. But I expect to hear from you young scholars what Heidegger is really saying. I wouldn't be at all surprised if he turned out to be the father of a new and deeper environmentalism.

Toward a Philosophy of Nature — The Bases for an Ecological Ethics

Murray Bookchin

Murray Bookchin: A pioneer in eco-philosophy, Dr. Bookchin's writings date from the early 1950's. His books include The Synthetic Environment*(1962),* The Limits of the City*(1973),* Toward An Ecological Society*(1980) and* The Ecology of Freedom*(1982) as well as other works on social theory. Retired Professor of Social Ecology from Ramapo College of New Jersey, and Rector Emeritus of the Institute for Social Ecology at Plainfield, Vermont, Bookchin currently lives in Burlington, Vermont where he is an avid gardener between lecture tours and literary commitments.*

PART I

Few philosophical areas have acquired the social relevancy in recent years that has been achieved by nature philosophy and its ethical implications. The literature on the subject has reached truly impressive proportions and collected a very sizable public following. Gregory Bateson's *Mind and Nature,* Arthur Koestler's *Janus,* Fritijof Capra's *Turning Point,* Leshan and Margenau's *Einstein's Space and Van Gogh's Sky,* Theodore Roszak's *Person/Planet,* and such insightful surveys as Carolyn Merchant's *The Death of Nature* and Morris Berman's *The Re-Enchantment of the World,* to cite only a few, have spilled over the traditional academic confines to which they would have been limited a generation ago, that is, if we are to assume that they could have flourished in such a format at all. Their readership currently exceeds any philosophical orientation that exists in the academy and its journals. A considerable segment of the literate public is now deeply occupied with the need for a philosophical interpretation of nature as a grounding for human behavior and social policy. In fact, it is fair to say that this public interest is comparable only to that which Darwinian evolutionary theory generated a century ago – and it is almost equally as disputatious and socially involved.

Current interest in nature's interface with society, however, differs basically from the ongoing dispute between notions of creation and evolution. Rather, this interest emerges from a deep public concern over the ecological dislocations that uniquely mark our era. Initially, this concern was largely technocratic and legalistic in its focus. It centered around pollution, resource depletion, demographic problems, urban sprawl, radiation, the increasing incidence of diseases like cancer, in short, the problems of conventional "environmentalism."[2] Environmental issues, in turn, were seen in strictly practical and limited terms that could be resolved by legislative action, public education, and personal example. Admittedly, such a strong environmental focus has not changed appreciably. It has merely shifted the centrality it gave to the issues with which it was preoccupied a decade or so ago and it has enlarged its scope to include more visibly political problems like nuclear weapons as well as nuclear power plants.

But the more philosophical literature that has emerged in recent years reflects a very significant dissatisfaction with such strictly issue-oriented approaches. This literature addresses itself to an entirely new body of concerns: the need, felt by a growing public, for an ecologically creative sensibility toward the environment – one that can serve in the highest ethical sense as a guide for human conduct and provide an awareness of humanity's "place in nature."[3] Perhaps the most important of these philosophical works deal with nature not merely as an environmental problematic. They advance a vision of the natural world that has been raised to the level of an inspirited metaphysical principle, yet without denying the significance of the environmental activism they seek to transcend. If the often narrow activism of the early and mid-seventies can be called the "politics" of environmentalism, the philosophical interests that are surfacing so prominently today can be called its "ethics," and certainly its social conscience.

Typically the academy is lagging behind this unique intellectualization of ecological problems, a theoretical turn that has been trying to satisfy a dire lack in popular perceptions of the current environmental crisis. The lag is all the more serious because the philosophical tradition that could greatly enrich this turn has been rendered needlessly technical or worse, reduced to mere historical and monographic memorabilia. Thus, much of what passes for nature philosophy today outside the campus tends to be wanting in roots in the western philosophical tradition, and such traditions as it does evoke have a strongly intuitional thrust.

Nor does the academy always add greater clarity to this ambiguous background when it intervenes with its own intellectual equipment and proclivities. Today, virtually *any* idea of a nature philosophy is burdened by a massive number of stultifying prejudices. The worst of these prejudices festers precisely in the academy where a conjunction of the words "nature" and "philosophy" automatically evokes images of anti-scientific "archaisms" and "anti-modernist" regressions to a static cosmological metaphysics. To speak frankly, the academic mind has been trained to view any nature philosophy as inimicable to critical and analytic thought. No less prejudicial in this regard are the "neo-Marxists," "post-Marxists," and emperical anarchists (for whom philosophy as such, short of Bertrand Russell's logical atomism, is sheer "theology") who uneasily regard all organicist theories as redolent of "dialectical materialism" and neo-fascist folk philosophies. If such prejudices are not dispelled – or, at least, explored more insightfully and critically – they will reinforce the quietism that has separated so many "leftist intellectuals" from "new social movements" which they all but damn as "defensive" and "co-optable."[4] They will foreclose any contact with a growing segment of the ecology movement that exhibits an exciting interest in self-reflection and a practice guided by theory in its broadest and deepest sense.

In any case, the literature exemplified by Bateson, Koestler, Capra, Roszak, and others will not conveniently disappear. Happily, if anything, this fascinating literature and its constituency is likely to grow. And the acolytes of the nature philosophies that are emerging today are as earnestly involved in speculation about nature and society as are feminist theorists with whom they increasingly overlap in such areas as anthropology, social theory, history, and in their fervent commitment to a new sensibility and culture. Either these concerns will be interwoven with the haunting tradition of a nature-oriented ontology and critically evaluated or they will have to be "reconstructed" along radically new lines. But the fact remains that a hitherto hidden naturalistic ethics has begun to surface that is no less real than the socially explicit "isms" that abound today. This ethics is seeking theoretical foundations as earnestly as the "founders" of socialism, anarchism, and liberalism did in their own day. It will not do for European and American academics to speckle this trend with learned name-droppings like "neo-Aristotelianism" or to closet it by evoking pedigrees as disparate as those created by Schelling, Driesch, Bergson, and Heidegger. Although contemporary excursions into nature philosophy – in *no* way to be confused with philosophies of science – require a broader philosophical grounding than they normally receive, they typically draw their nourishment from systems theory rather than the Greeks and Germans, and their hues are tinted by Asian rather than western cosmogonies. If such eclectic backgrounds seem discordant to our more schooled philosophical theorists, I would argue that they faced with the serious job of doing better rather than adding a new corps of prejudices to already existing ones. Whether one chooses to regard recent philosophical works on nature as a loss or a benefit to the ecology movement, this much is clear: if our schooled philosophical theorists decide to turn their backs to a rising theoretical trend within the "new social movements" which they have recently discovered growing underneath their feet, they will merely add to the distance that already separates them from some of the most important developments in contemporary society.

PART II

Before we can turn to the issues raised by the widely disparate theorists of the "new social movements," we must deal with a problem that unceasingly nags the acolytes of scientism and the spokesmen for philosophical "modernity." A troubling and eruptive "unconscious" plagues the philosophical "superego" of the academy and some of its self-professed radical theorists. When I speak of this "unconscious," I refer to what has been variously called the "Tradition" or, to use a more arrogant term, the "archaic" background that predates Enlightenment, indeed, modern, philosophy as a whole. The interest in Aristotle's *Metaphysics*, we are told, "remains perennial," so that it "does not flag or fail with the passing years, no matter how far the fashion of thought current at the moment may seem to wander from the confines of Aristotelian tradition."[5] But beyond this safely guarded work, the censor that acts like a screen on earlier philosophies seems to be remarkably secure. So little is permitted to filter pass the barrier raised by modern subjectivistic and scientistic orientations – and so unrecognizable are its forms – that the very origins of western philosophy have become a mystery to itself – in fact, a frightening specter like the primal nightmares of childhood that haunt the armored ego of the adult. Ironically, it has been left to Heidegger, among the very few who have been among us, to acknowledge that western philosophy has "origins," indeed, a fecund *arche*, that are worthy of exegesis, however much one may hesitate to follow all of his "woodpaths."[6] Even the word "ontology" wears a fearsome visage when it isn't qualified by adjectives such as "social," and the term "Being" has become so ghostly that it too has been subtly assimilated to subjectivistic strategies of dealing with reality, be they Hegel's or Heidegger's.

Limitations of space do not make it possible for me to enter into the problems which my remarks on these prejudices raise. But here, even some of the best theorists who are trying to intellectualize the ecological movement commit an error. Cognizant as they may be of the prejudices and the censoring mechanisms that have separated contemporary philosophy from its own history, they have dug their trenches poorly – with Descartes in mind rather than Kant. This is by no means an academic issue, nor is it strictly a philosophical one. The emphasis on Cartesian mechanism as the "original sin" that tainted the modern image of nature has been overstated for reasons that are more programmatic than theoretical. "Villanous" as Descartes may seem, there is a certain realpolitik involved, I suspect, that stresses him over Kant: a fear of surrendering the subjectivistic thrust that figures like Bateson give to systems theory and the quasi-religious character, a burgeoning transcendentalism, that has marked so much of contemporary "anti-mechanistic" thinking. Accordingly, philosophical theories of nature and the effort to derive an objective ecological ethics from them still exist in a false light and the "epistemological turn" which Kant finally gave to western philosophy has left pre-Kantian and ecological interpretations of nature as ambiguous as they are in the academy itself.

Presocratic philosophy, as Heidegger reminds us, is not the "dead dog" that it has become in conventional philosophy and the figures of Plato, Aristotle, and their offshoots, not to mention the festival of pantheism that burst forth with Bruno, remain as "perennial" and unflagging as Aristotle's *Metaphysics*. I am not concerned, for the present, with the *specific* speculations that pre-Kantian philosophies advanced – particularly, insofar as this concerns the Presocratics – but with the intentions and the kind of unities they tried to foster. Their message is easily neutralized if we follow Aristotle's and Hegel's track of dealing with them merely as *physiologoi*, boxing them into a tradition that simply yields an Aristotelian material and formal cause. What is important, as Gregory Vlastos has so admirably emphasized, is that they were the authentic and perhaps the least intellectually entangled voices of an objectivity permeated by ethics. Indeed, in contrast to the "naturalism" that became so fashionable in American academies during the thirties and forties, the unifying features of the Ionian, Eleatic, Herakitean, and Pythagorean trends are precisely the conviction that the universe has in some sense a moral character irrespective of human purposes – a proposition so alien to the post-Kantian era that it is dismissed as "archaic" and "teleological" almost as a knee-jerk reaction.[7]

Yet one cannot simply dismiss the matrix in which such great themes as Being, Form, Motion, and Causality were melded with moral meaning, an issue so crucial to ecological thought today. If one has the largeness of mind to ignore the specific accounts that each of the Presocratics used to explain the *arche* of the world, they can be conceived as following out the logic of a theme that permeates speculative philosophy to this very day. To *know* implies that we live in orderly world – an *intelligible* world – hence one that lends itself to rational interpretation because it *is* rational. However intuitively or consciously the case may be, this sequence of ideas about reality underpins the Hellenic notion of *nous* – of mind – that is either constitutive of the world or inheres in it. Neither so remote a figure as Thales nor so crucially a modern one as Hegel was to shed this essential orientation in which any philosophy and science is rooted, whether in part or in whole. For Laurence J. Henderson, in his immensely influential work, *The Fitness of the Environment* (1912), to cite an interesting case in point, the "Idea of purpose and order are among the first concepts regarding their environment which appear, a vague anticipation of philosophy and science, in the minds of men." For Henderson, it was the "advent of modern science" that validates universal order in the form of natural law. Darwin's hypothesis of natural selection, in turn, validates natural law "as the basis of purpose," specifically the "new scientific concept of fitness," thereby rescuing speculative thought from the "dogma of final causes . . ."[8]

What is important in Henderson's remarks is the *analytic context* that this orientation toward the world's intelligibility implies, not the specific answers that were to flesh it out and clothe it. Hellenic thought was especially unique because its flesh and cloth were pointedly moral *insofar as the world was rational,* i.e., that rationality and intelligibility were *equivalent* to morality. Hence, precisely because their attempt to provide an account of the world was explanatory, it was *meaningful*. It did not go halfway into the problem that order raised; it tried to resolve it to the full. Even such accounts of the *arche* of the world, (more precisely, its active substance) such as water (archaically, "blood" and "kinship"?), the "unbounded," or "aer" (which historians of Greek philosophy now regard as "soul," the "breath of life"), is itself redolent with meaning. It is this *sense* of reality as pregnant, fecund, and immanently self-elaborating that still provides the coordinates for an ecological philosophy – one that has been all too cheaply obscured by abjuring their "naivete," their "ontological need" (to use one of Adorno's less happy phrases), and their "monism."

And lest we place too much emphasis on their "naivete," let us bear we in mind that the Pythagorean *arche* – form – with the Pythagorean ideal of limit, of *kosmos* that combines order with beauty, and *krasis* as equilibrium have yet to find their match in the work of most ecological theorists. The remarkable, almost hypnotic richness of Greek philosophical terms which Heidegger has so superbly elaborated, has its origins in the Pythagorean community, a community that united within itself a commitment to a shared kinship between its members and the world. The term "holism" becomes meaningless when it is divested of the notion of form – a geometric notion that adds the formal concept of *arrangement* to the numerical notion of sum. Even more incisively than many ecological theorists today – who have yet to look beyond the disastrous plundering of the earth to its even more disastrous simplification – the image of form as the expression of the "good" and the "beautiful" renders virtue cosmically immanent. Philosophy, which Pythagoras regarded as a practice, thus structured rationality *qua* morality into a way of life, a form of ecological sociation or ecocommunity that has yet to be equalled in its level of consciousness by any phenomenon we know in our society.

But still more radically, the Presocratics were to anchor their interpretation of nature in the notion of *isonomia* – equality – a term that actually *guides* us to the "roots" or "elements," not one that is their predicate. With a sophistication that belies the image of the Presocratics as "primitive rationalists" who happily bridge the gap between the mythopoeic world and Plato and Aristotle, Presocratic thought is *consciously* infused by a notion of cosmic justice – *dikaisyne* – that reaches beyond social and personal issues to nature. Its counterpart, *adikiaisyne*, that marks every transgression of the law of the measure and the *peras* or limit of things and relationships, demands reparation and the restoration of harmony. From Anaximander to Empedokles, we encounter the need for a thorough-going respect for a ubiquitous principle of equality – indeed, a principle so consciously held that Alcmaeon was to use the term "monarchy" with opprobrium to characterize the "mastery" or "supremacy" of one cosmic power over another. *Krasis*, it is worth noting, is not the mechanical equipoise of contrasting powers but, more organically, their *blending*, and in the sequence of phenomena (initially, in Greek medical theory), their rotation. "As in the democratic polis 'the demos rules by turn,' [Euripides' phrase in the *Supplicants* – M.B.] so the hot could prevail in summer to the justice of the cold, if the latter had its turn in winter," Vlastos observes. "And if a similar and concurrent cycle of successive supremacy could be assumed to hold among the powers of the human body, then the krasis of man and nature would be perfect."[9]

This "elegant tissue of assumptions" is thoroughly naturalized by Empedocles whose concept of "roots" as distinguished from "elements," undifferentiated "Being," and "atoms" vastly enlarges the implicit Hellenic notion of an immanently generative nature to a point that we do not find even in Aristotle.[10] The "roots," as Cornford tells us, are "equal in status or lot"; they rotate their "rule" with their own unique "honor," for in no case is the universe a "monarchy" and none of its powers can claim even the primogeniture that Thales was to give to fluidity and Anaximenes to soul.[11] In this radical concept of rotation, we may well have the clue to the static nature of the more conservative Greek cosmologies, much as Plato's concept of forms with their fixidity and eternality aborted theories of evolution that certain Greek democrats were known to advance.

Taken by itself, the reworking of *isonomia*, *dikaisyne*, and *krasis* into statically transcendental concepts from more dynamically immanent and self-regulative Presocratic notions of nature is a regression. It is a regression that philosophy itself was to never fully overcome. Neither Plato nor Aristotle were prepared to follow Alcmaeon, among others, in leveling "the ancient inequalities which had been fixed by religious tradition."[12] Quite to the contrary, what we call "classical" Hellenic thought was to use rationalism, morality, and its own cosmological ideas to strengthen these ancient inequalities or to rear them in new forms. For the Presocratics, "Justice is not longer inscrutable moira, imposed by arbitrary forces with incalculable effect. Nor is the goddess Dike, moral and rational enough, but frail and unreliable." Unlike Hesiod's Dike, she is one with nature itself who "could no more leave the earth than the earth could leave its place in the firmament."[13]

Nature, in effect, became a commonwealth, a *polis*, whose *isonomia* effaced the "distinction between two grades of being – divine and mortal, lordly and subservient, noble and mean, of higher and lower honour. It was the ending of these distinctions that made nature autonomous and *therefore* completely and unexceptionally 'just.' Given a society of equals, it was assumed, justice was sure to follow, for none would have the power to dominate the rest. This assumption . . . had a strictly physical sense. It was accepted not as a political dogma but as a theorem in physical inquiry. It is, none the less, remarkable evidence of the confidence which the great age of Greek democracy possessed in the validity of the democratic idea – a confidence so robust that it survived translation into the first principles of cosmology and medical theory."[14]

Nature philosophy, I hold, was not to recover this egalitarian and reparative thrust – and it simply becomes silly to emphasize that Presocratic thought was riddled by demonstrably false "archaisms" when, in fact, we are exploring its *orientation*, not its "scientific" merits. Partly because we have been closed to this orientation by a historic screening of philosophic strategies which favored hierarchy over equality, transcendentalism over a self-regulative immanence, a Dike and Moira shaped by aristocratic concepts over a justice and destiny within the world of nature; partly, too, because nature philosophy in the classical era was to divide nature itself into "two grades of being" – a presiding *logos* or *nous* on one level over a degraded, irrational, and chaotic "matter" on the other – that nature philosophy lost not only its grounding in *isonomia* but its sense of ecological meaning. Taken as a whole, materialism did not overcome this bifurcation; it either tried to heal it or reverse the priorities given to these grades. "Cosmic equality lost its importance," Vlastos observes, "for cosmic justice no longer makes sense. Justice is now a human device; it applies solely to the acts and relations of conscious beings." And precisely this kind of justice becomes merely "the form which the immanent order of nature achieves in the mind and works of man. Justice is natural, but nature is not just."[15]

PART III

To join the issue of an ecologically oriented nature philosophy in this light rather than pose it as a simple contest between mechanism and some form of organicism helps us to understand the wayward fortunes of philosophy itself and the limits it imposed on a nature philosophy and an ecological ethics. We know that the founders of modern science – Copernicus, Kepler, and perhaps even Galileo – were raging Pythagoreans. What early Renaissance thought and science rescued from the ancients was form, not *isonomia*, as well as the shared premise of all speculative reason that nature was an intelligible *kosmos* and, intuitively perhaps, an intelligent one. Descartes, it must be said in all fairness, never gave us the ground to challenge this conceptual framework. He merely gave it a mechanical form – alluring and subversive for its time. Ironically, it was Kant – a near-Jacobin – who was to make the most significant turn in western philosophy with his "Copernican revolution," which totally denatures nature in the Presocratic sense by removing the material "grade of being" as such. Things-in-themselves simply cease to be "things"; they are merely acknowledgements that one "grade of being" has effectively ceased to exist for cognitive purposes. With Kant we are thus alone with our own logos, more precisely, our subjectivity. Jaspers sums the issue up incisively when he declares: "Kant does not, like all earlier philosophers, investigate objects; what he inquires into is our knowledge of objects. He provides no doctrine of the metaphysical world, but a critique of the reason that aspires to know it. He gives no doctrine of being as something objectively known, but an elucidation of existence as the situation of our consciousness. Or, in his own words, he provides no 'doctrine,' but a 'propaedeutics.' " Accordingly, Kantian categories have objective validity insofar as they remain within the limits of possible experience. "Metaphysics in the sense of objective knowledge of the supersensible or as ontology, which teaches being as whole, is impossible."[16]

Liberating as this may be from an absolutist empiricism that renders its own experience into a world of pure being, Kant does not free us from absolutes, notably his own sweeping intellectualization of objectivity. Despite his acknowledgement of a noumenal world that is both "supersensible" or "unknowable" and constitutes the originating source of the perceptions which the Kantian categories synthesize into authentic knowledge, Kant opened the way to an epistemological focus on *systems of knowledge* rather than *systems of facts*. More precisely, facticity itself was absorbed within systems of knowledge and the Greek *onta*, the "really existing things," were completely displaced by *episteme*, our "knowledge" of the now "unknowable" *onta*. Hegel was to poke fun at this patent contradiction – the knowing of an "unknowable" – but a veritable trench was drawn around philosophy which excluded nature as ontology and rendered thought into Being. Even in rejecting the dualism which this "agnostic" and essentially skeptical outlook created, Kant's "epistemological turn" in philosophy became absolutized. Whether we turn to Husserl or to Heidegger – to the process of *epoché* that brackets out the natural world in order to establish the logical necessity on which it will ultimately hang or to *Dasein* as the human existent and royal road to Being – reality is distilled into intellectuality and the *formalizations* of the human mind become the exclusive point of entry into Being. Indeed, these formalizations become Being itself, so that one can call Heidegger's philosophical strategy "ontological" – or, for that matter, Husserl's as well.

What is lost in this development are the *onta* that alone can provide the underpinnings of a nature philosophy, one that we are now obliged to distinguish from a philosophy of the "nature" of knowing. Even Hegel, who heaped scorn on the thing-in-itself – whose very thinghood by definition requires determinations and, "in fact," bears the imprint of the Kantian categories – ends in the very subjectivity that forms the ambiotic fluid of the Absolute. After the toil of Spirit, object and subject finally come to rest in Mind – in knowledge as self-knowing in all its totality. It is not for sentimental reasons that Hegel's *Encyclopedia* ends with a quotation from Aristotle that exults "thought [that] thinks on itself because it shares the notion of the object of thought."[17]

Ecological philosophy has not escaped from this Kantian trench. Indeed, it is a captive within it without even knowing so. Bateson, the most widely read of its gurus, is almost totally subjective in his interpretation of the notorious Mind-Nature relationship. In trying to properly "build the bridge" between "form and substance," he emphasizes only too correctly that western science began with the "wrong half" of the chasm, notably with an atomistic materialism. Bateson thus renders himself particularly attractive to ecologically oriented constituencies when he supplants matter by mind and cojoins fact (whatever that would mean) with value. Put quite systematically, any interrelational system is "Mind" and hence, in this sense, subjective – a notion that feeds into quasi-supernaturalistic visions of reality (generally Eastern in origin) that curiously tend to transcend the natural world rather than explain it. Batesonian mentalism, in turn, is nourished by the cybernetic idea that every perception must be seen as a system, not as isolates, or as Bateson puts it, "atomies." Epistemologically, this means that the differentia that form an aggregate of interacting parts are not spatial, temporal, or substantial; they are relational. The interaction that exists between a "subject" and an "object" is a kind of unit system that exists within ever-larger systems, be they communities, societies, the planet, solar system, and, ultimately, the universe. It is precisely these systems that Bateson designates as "Minds," more precisely, a *hierarchy* of "Minds" in a manner very similar to Koestler's "Holarchy," with its sub-levels of "holons" that reach from sub-atomic particles, through atoms, molecules, organelles, cells, tissues, organs up to living organisms with their own *scala natura*. That "Mind is empty; it is no-thing," means literally that it is no *thing* at all. Hence, *only* "ideas are immanent, embodied in their examples. And the example [that is, the material embodiments of ideas – M.B.] are, again, no-things. The claw, *as an example*, is not a *Ding an sich*; it precisely *not* the *'thing in itself.'* Rather it is what mind makes of it, an *example* of something or other."[18]

These remarks should not be taken as merely a subjectivist variant of Kantianism; they are more: a denial of thinghood *as such*. What Bateson is telling us is that context fixes meaning, a view that is not very new if one knows anything about Whitehead. But what renders Bateson a true offspring of the "epistemological turn" in western philosophy is his claim that "All experience is subjective . . . our brains make the images that we think we 'perceive.'" Indeed "occidental culture" lives under the "illusion" that its "visual image of the external world" has ontological reality. Accordingly, "mental striving" as an inherent characteristic of "the smallest atomies," a view which Bateson imputes to Teilhard de Chardin, introduces "the supernatural by the backdoor."[19] While this criticism may be valid insofar as it applies to Teilhard de Chardin, an acknowledged theologian, Bateson also happens to dismiss the very idea of ontological properties as such – a notion so essential to Enlightenment thinkers like Diderot – even as he smuggles them in for systems. What is actually an argument against "atomies" acquires the appearance of an argument against presuppositions. Actually, Bateson's own view is overloaded with presuppositions as, for example, his claim that his whole thesis "rests on the *premise* that mental function is immanent in the interaction of differentiated 'parts.' "[20] It is irrelevant whether this premise is valid or not. What counts, for exegetical purposes, is that is no less a presupposition than any advanced by Teilhard de Chardin.

Cybernetics, in turn, is uncritically taken for granted. That it could be regarded as another form of mechanism – electronic rather than mechanical – seems to have eluded most of its acolytes. Feedback loops are as "mechanistic" as flywheels, however different their physics may be, and the tendency for cyberneticians to accept a reductionism based on energy rather than matter retains the same logic that guided mechanical thinking in Newton's day. Odum's ecological cybernetics with its tunnel vision that can only perceive the flow of calories through an ecosystem is as shallow philosophically as it is useful practically within its own narrow limits. But among the more mystical acolytes of cybernetics, this energetics has taken the form of spiritual mechanism that eerily parallels all the failing of materialist mechanism without the latter's contact with reality. A deadening vocabulary of "information," "inputs," "outputs," "feedback," and "energy" has replaced such once-living terms like "knowledge," "dialogue," "explanation," "wisdom," and "vitality," a substitution that has occurred in blissful ignorance that cybernetic terminology was largely born with wartime research into radar and servo-mechanisms for military guidance systems.[21]

It is this larger cybernetic outlook, not merely Bateson's mentalism, that is of greater interest in the scheme of contemporary ideas. Just as the Presocratic notion of *isonomia* is inseparable in Vlastos' view from Hellenic notions of the democratic *polis,* so critics of Bateson's view and cybernetics generally have been quick to point out the authoritarian features of hierarchies of "Mind" and hierarchies of "holons" (Koestler).[22] Although Koestler has been acutely conscious of this problem in his own notion of "holarchy," Bateson, if anything, is given to examples that exploit it. Characteristically, Bateson, in describing "an alternating ladder of calibration and feedback up to larger and larger spheres of relevance and more and more abstract information and wider decision," emphasizes that "within the system of police and law enforcement, and indeed in all hierarchies, it is most undesirable to have direct contact between levels that are nonconsecutive. It is not good for the total organization to have a pipeline of communication between the driver of the automobile [who is ticketed for violating a speed-limit – M.B.] and the state police chief. Such communication is bad for the morale of the police force. Nor is it desirable for the policeman to have direct access to the legislature, which would undermine the authority of the police chef . . . In legal and administrative systems, such jumping of logical levels is called *ex post facto* legislation. In families, the analogous errors are called *double binds.* In genetics, the Weissmannian barrier which prevents the inheritance of acquired characteristics seems to prevent disasters of this kind in nature. To permit direct influence from somatic state to genetic structure might destroy the hierarchy of organization within the creature."[23]

This is sociobiology with a vengeance. Even such outstanding founders of system theory as Ludwig von Bertalanffy have been quick to distinguish general system theory from cybernetics for remarkably outspoken political and psychological reasons, particularly in the case of homeostasis. "The robot model or principle or reactivity . . . entails the *equilibrium theory* of behavior," Bertalanffy observes. "The natural state of the organism is that of rest. Every stimulus is a disturbance of equilibrium; behavioral response, there, is its re-establishment; it is *homeostasis, gratification of needs* or *relaxation of tensions.* The needs are essentially biological, pre-eminently hunger and sex. Again it follows that the behavior of animals such as rats, cats and monkeys provides the necessary bases for interpretation and control of human behavior; what appears to be special in man is secondary and ultimately to be reduced to biological drives and primary needs."[24]

Unfortunately, Bertalanffy's substitution of a generalized system theory for "Cartesian mechanism," "one-way causality," and "unorganized complexity," to use his own words, hardly provides us with a solution to the problems raised by cybernetic mechanism. Ultimately, the thinking in both cases is very much alike. A general system theory based on worldview of "organized complexity" is essentially a cybernetic system that is "open" rather than "closed." To use Bertalanffy's own words against him, a general system theory "is still mechanistic in the sense that it presupposes a 'mechanism,' that is, structural arrangements." Although it is quite true that "In behavioral parlance, the cybernetic model is the familiar S-R [stimulus-response – M.B.] . . . scheme" which simply replaces *"linear causality"* by *"circular causality* by way of the feed-back loop," the claims advanced by a general system theory of "multivariable interaction, maintenance of wholes in the construction of component parts, multilevel organization of ever-higher order, differentiation, centralization, progressive mechanization, steering and trigger causality, regulation toward higher organization, teleology and goal directedness in various forms and ways, etc." are geneally more programmatic than real and incorporate some of the most authoritarian as well as mechanistic attributes of cybernetics. That the "elaboration of this programs has only just begun . . . and is beset with difficulties" is an understatement.[25] To my knowledge, the only "breakthrough" that lends any credibility to Bertalanffy's sweeping claims for the explanatory potential of an "open system" in general system theory is Ilya Prigogine's mathematical elaboration of the organizing role of positive feedback.[26] Picked up by writers like Erich Jantsch and Fritijof Capra to deal with the problem of *development* – a solution to which is so markedly absent in system theory as a whole – Prigogine's work essentially utilizes the symmetry-breaking effects of positive feedback (more bluntly, disorder) as a means for creating "order" at various levels of organization.[27] This is certainly an original strategy. But valuable as it may be within the realm of system theory itself and particularly in its applications to chemistry, the idea itself is not very new. As to the reasons why spontaneous structuration occurs, they are no less "mechanistic" than Bateson's ladder of "Minds" or Koestler's hierarchy of "holons." In all of these cases, we are no closer to an explanation of *why* order and complexity emerge from other degrees of order and simplicity than why order emerges from the "disorder" produced by positive feedback.

The issue of development, specifically, evolution is crucial. Certainly none of the theories I have cited explains why one "level of organization" supersedes or incorporates another one. At best, each one tries to tell us *how*, and even in this respect, in a woefully incomplete manner. What is perhaps even more important, it is not clear whether cybernetics and general system theory can extend beyond mere *interaction* as distinguished from authentic *development*. Bateson's stochastic strategy for "explaining" sequence, for example, merely correlates random genetic mutations (worse, point mutations which are piecemeal as well as random) with a "selective process" that is itself remarkably passive. Natural selection merely tells us that the fittest survive environmental changes. If this correlation is all we know about evolutionary development —notably, the circular thesis that amidst a flurry of utterly random mutations, organisms capable of surviving are the "fittest" to survive—than we know very little about evolution indeed—and certainly, we have no "system" or "Mind" other than mere interaction that *explains* it. Indeed, not every "interaction" can be construed as *relationship* unless it is meaningful. To call the mere physical fact that one human being stumbles over another "intersubjectivity," for example, would be to degrade the very meaning of the word "subjective." All that has happened in this instance is that the encounter of one body with another has produced a form of physical contact. The "interaction" becomes "intersubjective" when the two persons address each other—possibly with friendly recognition, possibly with expletives, possibly even with blows. Moreover, I cannot emphasize too strongly in view of recent "formalizations" of even radical social theories, that our attempt to understand this "interaction" in all its possible forms and meanings requires that we know the social and psychological context in which it occurred, that is to say, *the history or dialectic*, however trivial, that lies buried within the "intersubjectivity" that results from the "interaction."

PART IV

Ultimately, there is no way to deal internally with cybernetic or system theory approaches to ecological issues, although we can certainly criticize their abuse of the term "hierarchy"—a strictly *social* term—as a substitute for *degrees* of complexity and organization. Like Kantian and neo-Kantian philosophies, they are basically self-sufficient and self-enclosed. While Kant's conclusions may not follow completely from his premises, the fact remains that his very errors have served as correctives for his successors. Put in the language of system theory, Kantianism and its subjective sequalae are sufficiently closed so that their own errors become the self-corrective source of perpetuating Kant's "Copernican revolution." That this turn has greatly enlarged philosophical thought is hardly arguable. The elaboration of an epistemology and the introduction of the subject as both observer and participant in cohering knowledge and reality have filled a major lacuna in western philosophy.

What is definitely arguable, however, are the imperial claims that this subjectivism advances, the totalization of reality and the arrogant exclusivity it has staked out for itself. Hegel's sniping against Kant, while indubitably shrewd, did not attentuate these imperial claims; indeed, to some degree, it played a corrective function for neo-Kantians or later generations. The real strength of Hegel's case against Kantianism in all its forms consists of the *alternative pathway* he opened up with the *Phenomenology of Spirit* and with his own phenomenological strategy, which is always part of his subject. To use Hegel's own description of this strategy: Insofar as the *Phenomenology* "has only phenomenal knowledge for its object, this exposition seems not to be Science, free and self-moving in its own peculiar shape; yet from this standpoint it can be regarded as the path of the natural consciousness which presses forward to true knowledge; or as the way of the Soul which journeys through the series of its own configurations as though they were the stations appointed for it by its own nature, so that it may purify itself for the life of the Spirit, and achieve finally, through a completed experience of itself, the awareness of what it really is in itself." This "pressing forward" is immanent to true knowledge, for short of finding its goal, "no satisfaction is to be found at any of the stations along the way."[28]

I share with Lukács, in contrast to the academic fluff who have vitiated Hegel's strong "reality principle," Engels' view that the *Phenomenology* may be called "a parallel of the embryology and the paleontology of the mind, a development of individual consciousness through its different stages, set in the form of an abbreviated reproduction of the stages through which the consciousness of man has passed in the course of history."[29] Like formalizations of Marx's *Capital*, an endeavor that goes back to the 1920s and a debate in which I participated in the "night school" world of socialism (to use Frederic Jameson's unsavory phrase) in the early 1940s, the formalization of the *Phenomenology* surrenders Hegel to Kantianism in the very name of determining its "authenticity." It is "no correspondence theory" of knowledge to acknowledge the remarkable extent to which the self-movement of consciousness in the *Phenomenology* so largely parallels the self-movement of consciousness in historical reality. What is important, however, is not the issue of formalistic distillations of Hegel's *Phenomenology* — which raises issues that may even go beyond Hegel's own intentions — but the extent to which the strategy is captive to natural reality and the ethical universe it opens for ecology. Once we grant that immanent critiques of system theories and subjectivistic approaches to nature are simply irrelevant if they fail to challenge the premises of the latter and establish alternative premises to them, it is incumbent upon us to formulate new premises that provide coherence and meaning to natural evolution. To put it bluntly: we are obliged to choose our premises carefully and adequately so that they yield coherence and meaning to reality, not rear the weary myth of a "presuppositionless philosophy" that constitutes an analytically elegant and deductively impeccable nature philosophy distilled of all ecological relevance and contact with reality. The truth or falsity of a nature philosophy lies in the truth or falsity of its unfolding in reality – in evolution as we are beginning to know it in nature today and as this natural evolution grades into social evolution and ethics.

The problem of choosing our presuppositions begins with our very *right* to claim *properties* for nature: to assume axiomatically that nature is defined by self-evident *attributes* as well as contexts. It is ironical that to make this statement immediately becomes problematical for a vast bouquet of academic philosophers, while it is certainly no problem among most scientists. The great Renaissance notion that "matter" and "motion" (or mass and energy, if you like) are basic *attributes* of nature, its most underlying "properties" (just as metabolism is a basic property of life) was a prevalent *scientific* assumption well into our own time and still remains so however much the definition of these terms and their relationships have been changed. It remained for Diderot in his extraordinary "D'Alembert's Dream" to propose the crucial trait of nature that transforms mere motion into development and directivity; the notion of *sensibilité*, an internal nisus, that is commonly translated as "sensitivity."[30] This immanent fecundity of "matter" – as distinguished from motion as mere change of place – scored a marked advance over the prevalent mechanism of LaMetrrie and, by common acknowledgement, anticipated 19th century theories of evolution, indeed, in my view, recent developments in biology. *Sensibilitié* implies an *active* concept of "matter" that yield increasing complexity from the atomic level to the brain. Continuity is preserved through this development without any reductionism; indeed, the *scala natura*, now dynamized by Diderot's avowed Heraklitean bias for flux, acquires its evolutionary center in a nisus for complexity, an *entelechia* that emerges from the *very nature, structure, and form of potentiality itself*, given varying degrees of the organization of "matter." It is from this *primacy* of the concept of potentiality and the actualization of the potentialities of various organisms, that the notion of *sensibilité* initiates its "journey" of self-actualization and emergent form. Diderot's holism, in turn, is one of the most conspicuous features of the "Dream," which by its very title, forewarns the readers of Diderot's candid sense of doubt, his own "likely story," given the limited scientific knowledge of his time. An organism achieves its unity and sense of direction from the contextual wholeness of which it is part, a wholeness that imparts directiveness to the organism and reciprocally receives directiveness from it. Apart from its systematic and mathematical treatment of feedback, there is little that cybernetics or system theory can add to this idea advanced by an authentic (and largely unacknowledged) genius who died almost two centuries ago. That Denis Diderot accepted the sensationalist theories of his time is certainly more understandable than the fact that so many "neo-" and "post-Marxists" have immersed themselves in communications theory. The active and directive "matter" which Diderot advanced with his notion of *sensibilité* marks a radical breach not only with Renaissance and Enlightenment mechanism but its relevance as "sensitivity," however metamorphic the terminology, is radically important for understanding current development in contemporary natural science.

If Hegel's phenomenological strategy is combined with Diderot's concept of *sensibilité* in "matter," we emerge with a fascinating possibility. Nature seems to be writing its own natural philosophy and ethics —not the logicians, positivists, neo-Kantians, and heirs of Galilean scientism. We are not alone in the universe, not even in the "emptiness" of space. Owing to what is a fairly recent revolution in astrophysics (possibly comparable only to the achievements of Copernicus and Kepler), the cosmos is opening itself up to us in new ways that call for an exhilarating speculative turn of mind and a more qualitative approach to natural phenomena. It is becoming increasingly tenable to suggest that the entire universe may be the cradle of life — not merely our own planet or planets like it. The formation of all the elements from hydrogen and helium, their combination into small molecules and later into self-forming macromolecules, and finally the organization of these macromolecules into the constituents of life and possibly mind follow a sequence that challenges Russell's image of humanity as an accidental spark in a meaningless void. The presence of complex organic molecules in the vast reaches of the universe are replacing the classical image of space as a void by an image of space as a restlessly active chemogenetic ground for an astonishing sequence of increasingly complex chemical compounds. Similarly, new theories about the formation of DNA modelled on the activity of crystalline replication, a notion advanced as early as 1944 by Edwin Schrödinger, suggest how genetic guidance and evolution might have emerged to form an interface betwen the inorganic and organic.[31]

The point is that we can no longer be satisfied with an inert "matter" that fortuitously collects into living substances. The universe bears witness to an ever-striving, developing — not merely moving — substance, whose most dynamic and creative attribute is its unceasing capacity for self-organization into increasingly complex forms. Nor can we remove form from its central place in this developmental and growth process, or function as an indispensable correlate of form. The orderly universe that makes science a possible project and its use of a highly concise logic — mathematics — meaningful presupposes the correlation of form with function. Life simply marks a *graded* development beyond the crucible of chemogenetic complexity we call the universe in which metabolism, added to development, establishes the existence of another elaboration of *sensibilité* – a trait we call symbiosis. Recent data support the view that Peter Kropotkin's mutualistic naturalism not only applies to relationships between species, but also applies morphologically – within and among complex cellular forms. As William Trager was to observe a decade ago in his ironical remarks about the "struggle for existence" and the "survival of the fittest": "few people realize that mutual cooperation between different kinds of organisms – symbiosis – is just as important, and that the 'fittest' may be the one that most helps another to survive."[32]

Indeed, the cellular structure of all multi-cellular organisms is testimony to precisely a symbiotic arrangement that renders complex forms of life – human as well as nonhuman – possible. The eukaryotic cell, to which I refer here, is a highly functional symbiotic arrangement of the less complex and more primal prokaryotes which evolved in an anaerobic world, long before our highly oxygenated atmosphere was formed. Owing largely to the work of Lynn Margulis, we have reason to believe that eukaryotic flagella derive from anaerobic spirochetes; mitochondria, from prokaryotic bacteria that were capable of respiration as well as fermentation; and plant chloroplasts, from "blue-green algae," which have recently been reclassified as cyanobacteria.[33] If it is true, as Manfred Eigen has argued, that evolution "appears to be an inevitable event, given the presence of certain matter with specified autocatalytic properties and under the maintenance of the finite (free) energy flow [that is, solar energy – M.B.] necessary to compensate for the steady production of energy," our very concept of "matter" too has to be radically revised.[34] The prospect that life and all its attributes are latent in "matter" as such, that biological evolution is rooted deeply in symbiosis or mutualism, suggests that what we call "matter" more properly can be characterized as active substance. Moreover, in a larger image of what we call "life," the Weissmannian barrier that conveniently separates genetic changes from somatic ones ceases to be meaningful from an ecological viewpoint, which sees life in its environment, not merely within its own skin, as it were. The traditional dualism between the living and non-living world, between the organism and its abiotic ecosystem, is being replaced by the more challenging notion that life "makes much of its own environment," to use Margulis' words. "Certain properties of the atmosphere, sediments, and hydrosphere are controlled by and for the biosphere." By comparing lifeless planets such as Mars and Venus with the Earth, Margulis notes that the high concentration of oxygen in our atmosphere is anomalous in contrast with the carbon dioxide worlds of the other planets. Moreover, "the concentration of oxygen in the Earth's atmosphere remains constant in the presence of nitrogen, methane, hydrogen, and other potential reactants." Life, in effect, exerts an active role in maintaining free oxygen molecules and their relative constancy in the earth's atmosphere. The uniqueness and anomalies of the Earth's atmosphere "are far from random." Much the same can be said for the temperature of the Earth's surface, the salinity of its oceans, and other seemingly random processes whose very stability seems to be a function of life on the planet. What we designate as the "selective" features of nature and Darwinian evolution become the products of the very life-forms, or at least, life itself whose "random" genetic changes they presumably filter out.[35]

Even the Modern Synthesis, to use Julian Huxley's term for the neo-Darwinian model of organic evolution in force since the early 1940s, has also been challenged as too narrow and perhaps too mechanistic in its outlook. The image of a slow pace of evolutionary change emerging from the interplay of small variations, which are "selected" for their "adaptability" to the "environment," is no longer as tenable as it seemed by the actual facts of the fossil record. Evolution seems to be more sporadic, marked by occasional changes of considerable rapidity, often delayed by long periods of stasis. This "Effect Hypothesis," advanced by Elizabeth Vrba, suggests that evolution seems to include an immanent striving, not merely random mutational changes in response to the filtering effect of external selective factors. As one observer notes, "Whereas species selection puts the forces of change on environmental conditions, the Effect Hypothesis looks to internal parameters that affect the rates of speciation and extinction."[35] Indeed, the notion of small, gradual point mutations (a theory that accords with the Victorian mentality of strictly fortuitous evolutionary changes, much like the operations of the 19th-century image of the marketplace) can be challenged on genetic ground alone. Not only a gene but a chromosome may be altered chemically and mechanically. Genetic changes may range from "simple" point mutations, through jumping genes and transposable elements, to major chromosomal rearrangements. Major morphological changes may thus result from *mosaics* of genetic changes, which raises the very intriguing question of a directiveness within genetic change itself, not simply a promiscuous and purely fortuitous randomness that is "selected" for its "fitness" by an environment largely created *by life itself*, not by forces exclusively external to it.

There is no mysticism or anthropocentrism involved in an ecological image of nature that monistically and ontologically grades natural history into social history without sacrificing the unity of each, nor is it a supernatural fallacy to derive the human brain from an actively chemogenetic universe that is self-forming, self-directive, and immanently entelechical. If Driesch has given the word "entelechy" a bad name, we would do well to remember that it derives from Aristotle, not from a confused neo-vitalism of recent times. The Presocratic idea of *isonomia* has its analogue in a shared continuity of evolution that is physically and biologically "egalitarian." Chemically, hydrogen is still the most abundant as well as the most primal element in the universe and micro-organisms, still highly archetypal in structure and function, render all complex forms of life possible. The *scala natura* of the living and non-living, each scaled to "dominate" the other or "obey" it is a fiction that projects human hierarchy on the world around it. It stands in flat defiance to a universal *isonomia* that renders the entire biotic world mutually heteronomous — a mosaic rather than a pyramid, a pattern rather than a ladder.

The fallacies of classical cosmology generally lie not in its ethical orientation but in its dualistic approach to nature. For all its emphasis on speculation at the expense of experimentation, ancient cosmology erred most when it tried to cojoin a self-organizing, fecund nature inherited from the Ionians with a vitalizing force alien to the natural world itself. Parmenides' Dike, like Bergson's *élan vital*, are substitutes for the self-organizing properties of nature, not motivating forces within nature that account for an ordered world. A latent dualism exists in monistic cosmologies that try to bring humanity and nature into ethical commonality — a *deus ex machina* that corrects imbalances either in a disequilibriated cosmos or in an irrational society. Truth wears an unseen crown in the form of God or Spirit, for nature can never be trusted to develop on its own spontaneous grounds, any more than the body politic bequeathed to us by "civilization" can be trusted to manage its own affairs.

These archaisms, with their theological nuances and their tightly formulated teleologies, have been justly viewed as socially reactionary traps. In fact, they tainted the works of Aristotle and Hegel as surely as they mesmerized the minds of the medieval Schoolmen. But the errors of classical nature philosophy lie not in its project of eliciting an ethics from nature, but in the spirit of domination that poisoned it from the start with a presiding, often authoritarian, Supernatural "arbiter" who weighed out and corrected the imbalances or "injustices" that erupted in nature. Hence the ancient gods were there all the time, particularly from Heraklitos onward; they had to be exorcised in order to render an ethical continuum between nature and humanity more meaningful and democratic. Tragically, late Renaissance thought was hardly more democratic than its antecedents, and neither Galileo in science nor Descartes in philosophy performed this much-needed act of surgery satisfactorily. They and their more recent heirs separated the domains of nature and mind, recreating deities of their own in the form of scientistic and epistemological biases that are no less tainted by domination than the classical traditions they demolished.

Today, we are faced with the possibility of permitting nature — not Dike, God, Spirit, or an *élan vital* — to open itself to us ethically on its own terms. Mutualism is an intrinsic good by virtue of its function in fostering the evolution of natural variety and complexity. We require no Dike on the one hand or canons of "scientific objectivity" on the other to affirm the role of community as a desideratum in nature and society. Similarly, freedom is an intrinsic good; its claims are validated by what Hans Jonas so perceptively called the "inwardness" of life forms, their "organic identity" and "adventure of form." The clearly visible effort, venture, indeed self-recognition, which every living being exercises in the course of "its precarious metabolic continuity" to preserve itself reveals — even in the most rudimentary of organisms — a sense of identity and selective activity which Jonas has very appropriately called evidence of "germinal freedom."[37]

"Open systems," "Minds," and "holons" may explain the disequilibria that *change* cybernetic and general systems, but we are invariably obliged to fall back on inherent attributes of substance, notably the motion, form, and *sensibilité* of "matter," to account for the directive development of nature toward complexity, specialization, and, I would add, a subjectivity we call "consciousness." This argument runs counter to every bias in philosophy, today, which would even ignore the fact of directiveness or personalize it with self-vitiating words like "purpose" and "goal" when we need talk of nothing more than tendencies that inhere in the organization of substance as potentialities. Nevertheless, contemporary science's greatest achievement is its growing evidence that the random, while it certainly exists, is secondary to a directive ordering principle, one that requires no greater intellectual justification than the fact of Being itself. Presupposition, here, does not mean an arbitrary choice of foundations. Whether a presupposition is valid or not must be tested against the real *dialectic* of facts — substance "free and self-moving in its own peculiar shape" — not the "atomies" of "data" and "statistical probabilities" adduced by empirical observation. On this score, contextualists like Whitehead and Bateson are quite sound in their claims that "facts" do not exist on their own, that they are always "relational" or more properly, *interactive*, to use Diderot's more germinal language. Admittedly, this approach to a nature philosophy is as self-sufficient and self-enclosed as the Kantian. But I have not faulted Kantian, neo-Kantian, or for that matter, cybernetic and positivistic theories for their internal unity and their impregnability to immanent criticism. My objection to them is that in their claims to universality, the presuppositions they make, do not provide an adequate framework for the natural history science is opening to us and the ethical implications that follow from it.

What I am asserting, in conclusion, is that our study of nature exhibits a self-evolving patterning, a "grain," so to speak that is implicitly ethical. Mutualism, freedom, and subjectivity are not strictly human values or concerns. They appear, however germinally, in larger cosmic or organic processes that require no Aristotelian God to motivate them, no Hegelian Spirit to vitalize them. If social ecology provides little more than a coherent focus to the unity of mutualism, freedom, and subjectivity as aspects of a cooperative society that is free of domination and guided by reflection and reason, it will remove the taints that blemished a naturalistic ethics almost from its inception. It will provide both humanity and nature with a common ethical voice. No longer would we need a Cartesian and Kantian dualism that leaves nature "mute" and mind isolated from the larger world of reality around it. We would see that the natural history of mind *is* mind itself – from the very *sensibilité* of the inorganic to the conceptual capacities of the human brain – not *sui generis* in a world that is external to it. To vitiate community, to arrest the spontaneity of a self-organizing reality toward ever-greater complexity and rationality as nature rendered self-consciousness, to deny the *isonomia* that underpins a sound ecological approach in contradistinction to a hierarchical one – these actions would cut across the grain of nature, deny our heritage in its evolutionary processes, and dissolve our legitimacy in the world of life. No less than this ethically rooted legitimation of humanity as nature rendered self-conscious would be at stake.

Mutualism, self-organization, freedom, and subjectivity, cohered by social ecology's principles of unity in diversity, spontaneity, and nonhierarchical relationships, are thus *ends that exist in themselves*. Aside from the ecological responsibilities they confer on our species as the self-reflexive voice of nature, they literally define us. Nature does not "exist" for us to use; it simply legitimates us and our uniqueness ecologically. Like the concept of Being, these principles of social ecology require no analyses but merely verification. They are the elements of an ethical ontology, not rules of a game that can be changed to suit one's personal needs and interests.

A society that cuts against the grain of this ontology raises the entire question of its very reality as a meaningful and rationality entity. "Civilization" has bequeathed us a vision of otherness as "polarization" and "defiance," and of organic "inwardness" as a perpetual "war" for self-identity. This vision threatens to subvert the ecological legitimation of humanity and the reality of society as a potentially rational dimension of the world around us. It is the agnostic world of the male warrior – the hunter whose killing of game has turned into a militarized predator who kills people. Trapped by the false perception of a nature that stands in perpetual opposition to our humanity, we have redefined human development itself to mean strife as a condition for pacification, control as a condition for consciousness, domination as a condition for freedom, and opposition as a condition for reconciliation. Within this implicitly self-destructive contest, we are building the ideological and physical walls that will almost certainly become a trap rather than a home within the world that nourishes us.

Yet an entirely different philosophical and social dispensation can be read from the concept of otherness and inwardness of life. Given a world that life itself has made conducive to evolution – indeed, benign, in view of a larger ecological vision of nature – we can formulate an ethics of complementarity that is rooted in variety rather than one that guards individual inwardness from a threatening, invasive otherness. Indeed, the inwardness of life can be seen as an expression of a dynamic and creative equilibrium, not as mere resistance to entropy and the terminus of all activity.

"It is not man's lapse into luxurious indolence that is to be feared, but the savage spread of the social under the mask of universal nature, the collective as a blind fury of activity," Adorno was to declare in the mid-forties when the Second World War came to a close. "The naive supposition of an unambiguous development toward increasing production [and one is tempted to add, the domination of nature – M.B.] is itself a piece of that bourgeois outlook which permits development in only one direction because, integrated into a totality, dominated by quantification, it is hostile to qualitative difference. . . . If uninhibited people are by no means the most agreeable or even the freest, a society free of its fetters might take thought that even the forces of production are not the deepest substratum of man, but represent his historical form adapted to the production of commodities. Perhaps the true society will grow tired of development and, out of freedom, leave possibilities unused, instead of storming under a confused compulsion to the conquest of strange stars. A mankind which no longer knows want will begin to have an inkling of the delusory, futile nature of all the arrangements hitherto made in order to escape want, which used wealth to reproduce want on a larger scale. Enjoyment itself would be affected, just as its present framework is inseparable from operating, planning, having one's way, subjugating. *Rien faire comme une betê*, lying on water and looking peacefully at the sky, 'being, nothing else, without any further definition and fulfillment,' might take the place of process, act, satisfaction, and so truly keep the promise of dialectical logic that it would culminate in its origins."[38]

Lest this seem too much like the Island of the Lotus-Eaters, bear well in mind the term "might," the expression of ambiguity that always marks the contrareity of Adorno's writings. As a testament against a "blind fury of activity," it is unerring in its insight into the deepest problems raised by human history and by our interpretation of natural history. Yet alluringly, there lies hidden within this faltering sense of promise the possibility that humanity could have found its sense of self-identity and individuation through ecological differentiation rather than hierarchical opposition; that the "I" could have formed itself around mutuality, with its wealth of uniqueness, rather than the commanding "lordship" with all its reversals that we find in Hegel's "master-slave" relationship. Indeed, the origins with which dialectical logic keeps faith in its return, rich in its wealth of differentiation, could have been more of a "whole" that we realize after looking over the blighted landscape of our history.[39] Here is the faith that social ecology keeps with the promise opened by Adorno and his colleagues in the Frankfurt School. In any case, "modernity," as we understand it today, may not be a "nullity," but it is certainly no garden of roses either.[40]

September 30, 1982

NOTES

[1]This article is a reconnoiter into an area that has occupied my thoughts for many years. It should be read as a tentative venture that requires fuller development and correction. It is obviously also limited by considerations of space and editorial needs.

[2]For my distinction between "environmentalism" and ecology – more precisely, social ecology – see my essay "Toward an Ecological Society" initially delivered as a lecture at the University of Michigan, Ann Arbor, in the spring of 1973. This essay was published during the same year in *Roots* and *WIN* magazine. It is now available as the leading essay in a collection of my 1970s writings, *Toward an Ecological Society* (Montreal, 1980) published by Black Rose Books.

[3]This phrase is taken, of course, from Max Scheler's *Man's Place in Nature*.

[4]The best example of this question is found in Jürgen Habermas' "New Social Movements," *Telos*, No. 49 (Summer, 1981).

[5]Joseph Owens' "Foreword" in Giovanni Reale, *The Concept of First Philosophy and the Unity of the "Metaphysics" in Aristotle (Albany, N.Y., 1980), p. xv.*

[6]Cf. Martin Heidegger, *Early Greek Thinking* (New York, 1975).

[7]Cf. Gregory Vlastos, "Equality and Justice in Early Greek Cosmologies" in *Studies in Presocratic Philosophy*, Vol. I, ed. Furley and Allen (New York, 1970), pp. 56-91.

[8]Laurence J. Henderson, *The Fitness of the Environment* (Boston, 1958), pp. 1, 5.

[9]Vlastos, *op. cit.*, p. 60. Heraklitos, the least democratic of the Presocratics, does not speak of *isonomia* but of the "One," which we can properly distinguish from the "Whole." This mystical thrust already prefigures neo-Platonism which was to emphasize the transcendental and the socially elitist elements in Greek philosophy.

[10]The words are Vlastos. For a discussion of its breakdown, see *ibid.* pp. 86-91.

[11]F.M. Cornford, *From Religion to Philosophy* (London, 1917), p. 64.

[12]Vlastos, *op. cit.*, pp. 59, 62, 63.

[13]*Ibid.*, p. 84.

[14]*Ibid.*, p. 85.

[15]*Ibid.*, pp. 90-91.

[16]Karl Jaspers, *Kant* (New York, 1957), pp. 50, 51.

[17]G.W.F. Hegel, *The Philosophy of Mind* (Oxford, 1971), p. 315.

[18] Gregory Bateson, *Mind and Nature* (New York, 1979), p. 11.

[19] *Ibid.*, pp. 31, 93.

[20] *Ibid.*, p. 93.

[21] I have explored the mechanistic aspects of cybernetics and system theory elsewhere. See my "Energy, 'Ecotechnology,' and Ecology" originally published in *Liberation*, February, 1975, and recently republished in my book *Toward an Ecological Society* (Montreal, 1980), pp. 87-96. This predictable inversion of a materialist mechanism into a spiritual mechanism has since been advanced by many writers, most notably Carol Merchant in *The Death of Nature* (San Francisco, 1980).

[22] Arthur Koestler, *Janus* (New York, 1978), pp. 30-34. Ironically, Koestler hypostasizes "hierarchy" in counterposition to reductionism even as the term haunts him because "it is loaded with military and ecclesiastical associations . . . [and] conveys the impression of a rigid authoritarian structure . . . ," a view I will certainly not dispute, despite Koestler's intellectual acrobatics in his effort to rescue the word as an expression of "flexibility and freedom." (p. 34)

[23] Bateson, *op. cit.*, p. 199. Here, Bateson exhibits the highly authoritarian character of his social outlook and pedigree which Morris Berman, an admirer of Bateson's work, so carefully explores in *The Re-Enchantment of the World* (Ithaca, N.Y., 1981), pp. 280-96. I disagree with Berman's view that an anarchic ecological society follows from Bateson's cybernetic approach. And this is precisely what I am trying to argue in my criticism of system theory generally – its lack of a truly qualitative and developmental approach.

[24] Ludwig von Bertalanffy, *Robots, Men and Minds* (New York, 1967), p. 9. This is a robust and gallant book by one of the finest and most educated minds among the founders of system theory and certainly the most sophisticated. Which does not spare Bertalanffy from being captive to the closed "feedback loops" he so earnestly tries to criticize and their authoritarian implications.

[25] *Ibid.*, pp. 67, 71.

[26] Gregoire Nicolis and Illya Prigogine, *Self-Organization in Nonequilibrium Systems* (New York, 1977).

[27] See Erich Jantsch, *The Self-Organizing Universe* (New York, 1980) and Fritijof Capra, *The Turning Point* (New York, 1982). Certainly Jantsch's account makes for excellent reading. But what is lacking in his overall approach is an organic one. Bertalanffy's criticism that cybernetics "presupposes a mechanism" (the words "structural arrangement" are rather maladroit) is precisely the issue.

[28] G.W.F. Hegel, *Phenomenology of Spirit* (Miller translation) (Oxford, 1977), p. 49.

[29] Quoted in George Lukács, *The Young Hegel* (Cambridge, 1966), p. 468. It is refreshing, today, to read Lukács' works with their healthy realism and "night school" concerns, compared with Frederic Jameson's "graduate seminar" world of term papers and PhD dissertations.

[30] See Denis Diderot, *Selected Writings* (Steward and Kemp translation) (New York, 1936), pp. 64-118. This version, by far the best translation of Diderot's works, captures the elegance and rich nuance of Diderot's writings – attributes that are often lost in other translations of his work into English.

[31] See Edwin Schrödinger, *What is Life? Mind and Matter* (New York, 1944). For a more detailed account of the new advances in astrophysics and biology, see my *Ecology of Freedom* (Palo Alto, 1982) from which a number of these passages, generally in modified form, are drawn.

[32] William Trager, *Symbiosis* (New York, 1970), p. vii.

[33] Lynn Margulis, *Symbiosis in Cell Evolution* (San Francisco, 1981).

[34] Manfred Eigen, "Molecular Self-Organization and the Early Stages of Evolution," *Quarterly Review of Biophysics*, Vol. 4, No. 2-3 (1971), p. 202.

[35] Margulis, *op. cit.*, pp. 348-49.

[36] Vrba cited in Robert Lewin, "Evolutionary Theory Under Fire," *Science*, Vol. 210, No. 1 (1980), p. 885.

[37] Hans Jonas, *The Phenomenon of Life* (New York, 1966), pp. 82, 90.

[38] Theodore Adorno, *Minima Moralia* (London, 1974), pp. 156-57.

[39] Increasingly, I have become convinced over the years that we broke our "faith" with the dialectic of society many millenia ago when a woman-oriented social development was displaced by a man-oriented one. I draw this sharp distinction between two very different societies advisedly – as distinguished by the magnificently transcendental term "society" *as such*. My own studies in anthropology and history, even in mammalian communities, have convinced me that the words "human society" are distinctly gender-laden, notably in the male's favor and what we call "society" is seen more through the eyes of the male fraternities that formed around the "men's house" in tribal communities than woman's sororal groups. Yet the two are distinctly separate and constitute two very different albeit co-existing societies. At some distant time before the threshold of history and currently, to some degree, among the debris of existing preliterate communities, there is compelling evidence that women formed their own unique social relationships, exclusive of males, in their workaday activities of domestic life. They grouped together as coworkers, mothers, sisters in a cultural sense, and administrators of their own sphere of life with their own technics and modes of expression, dances, and rituals that were as culturally rich, psychologically supportive, and spiritually nourishing as those which the men professed to form as warriors, comrades, hunters, and ultimately as a civil community. C.M. Turnbull's account of female lifeways and the cynicism of women toward male posturing and boastfulness among the Ituri forest pygmies (see the *Forest People*, 1969) and particularly the remarkable study by Yolanda and Robert F. Murphy of the Mundurucú Indians of the Amazon (see *Women of the Forest*, New York, 1974) are only the most noteworthy descriptions of the two gender-centered societies we so facilely group under the rubric of one society. Inasmuch as "human society" is always seen through the male's eyes, much as though social life is always seen through the eyes of the State, it is not surprising that we see "male dominance" at all times, both in the past as well as the present, and woman, following the imagery of the patricentric anthropology of Levi-Strauss, is viewed as a "commodity" for linking male fraternities through marriage. Even the Murphys fall prey to this prejudice despite the fact that all their evidence and many of their generalizations stand in flat defiance of their structuralist biases. Owing to the fact that males tend to be forceful in their behavior and tend to control the civil institutions of a community, it is assumed that their activities and social forms constitute the totality of society and that woman's domain is somehow marginal and heteronomous.

It is erroneous to say that we have never written a "history" of women or of the oppressed. We have done so only too well, particularly in the case of women – but we have done so only through the eyes of men or male ruling elites. We have written this history to obscure the fact that woman had her own society with its own integrity and citizenship – a female domestic society rooted in ecological differentiation, mutuality, and wholeness – only to be edged out by the male's civil society rooted in hierarchical opposition, rivalry, and one-sidedness. It is precisely the latter that is invariably described as "human society" and almost by definition submerges woman's world to one of its facets rather than a nexus of social relationships that exist in its own right. Yet however much we have tried to degrade woman as "gossipy," "flighty," "irrational," "inquisitive," and "emotional," her ancient sorority rises up to haunt our male garrison-world with its promise of an evolutionary pathway that might have yielded a truly pacified, mutualistic, egalitarian society in which neither men nor women would preside over each other, nor human society preside over nature. Only by evoking that memory and learning from its remanent forms – that is, by stepping back anthropologically to the point where the road divided between woman's submission to man's dominance – that we can hope to pick up earlier threads of a social pathway whose further elaboration and development can spare us from mutual annihilation.

[40]The terms "modernity" as "nullity" are not mine. They were written by an admirer of Jürgen Habermas, the self-anointed "heir" of the Frankfurt School, in criticism of my presumed "eschatological" and "messianic" views. For my reply, which summarizes most of the issues involved, see my "Finding the Subject," *Telos*, No. 52 (Summer, 1982), pp. 78-98.

Nature and Culture in Japan

Allan G. Grapard

Allen Grapard: One of the world's foremost scholars of Japanese Buddhism, currently directing an international research program on the symbolism of the lotus in Japanese culture. He is an East-West Center Research Fellow, former Professor of Asian Religions at Cornell, and Social Science Research Council Fellow in Kyoto, where he has lived for many years. He has written two ground-breaking articles in the History of Religions *journal at the University of Chicago: "Flying Mountains and Walkers of Emptiness" (1982), and "Japan's Ignored Cultural Revolution"(1984). In addition, he has two major forthcoming works:* Introduction to the Kasuga Cult, *and* The Final Truth of the Three Teachings, *this latter work for the Paris National Radio Station publications series.*

The history of massive environmental abuse which has marked the West has also been that of Japan at least since its opening to the West in 1868. Having imported indiscriminately most of the West's industrial and economic practices, as well as quite a few "cultural" assumptions which accompany these practices, the Japanese have come to face very much the same problems the West is confronted with. One of the positive results of the mindless destructions which we can see occur in our environment has been the appearance of systematic thinking on the relationships between nature and culture, as evidenced by the ecological movements in Europe and in the United States. It is interesting to note that the evolution of ecophilosophy in the West indebts itself consciously to Asian systems of thought and practice, perhaps because some Western scholars of Asian traditions want to see in these the philosophical possibility that anthropocentrism is a fallacy. The deterioration of cities, the decrease in the quality of life, and the stunningly swift disappearance of wildlife and wilderness areas appear to be as many proofs of the idea that these developments may have resulted from particular culturally-grounded attitudes toward nature, or from particular socio-economic processes against which only a marriage between philosophy and politics could fight successfully.

The following pages are an attempt to suggest that the Japanese cultural tradition hides in its deepest recesses a vast storehouse of notions and practices which may be helpful in establishing a culturally-grounded ecophilosophy. The method used in this short article is simple and could be developed in complexity if it is agreed that it is useful in evidencing the presence of environmental ethics in the tradition: having gone through a large number of texts belonging to the philosophico-religious traditions of Japan, and having witnessed a number of ritual practices which could be interpreted in the light of these documents and in the light of contemporary Western thinking on the topic of deep ecology, we have decided to propose a few reflections on what could be called "base-models" of environmental ethics.

NATURE AND CULTURE IN JAPANESE MYTHOLOGY

It is useful to investigate Japanese mythology in a search for base-models not only because one finds there some of the earliest written statements about the environment, but also because this mythology is still alive in rituals performed in many Shintō Shrines and is, therefore, of some relevance. These rituals and their accompanying symbols are thus still close to the mind, though it may be said that their presence is partly hidden from consciousness and belongs to the world of the unconscious, of dreams, and of images.

According to the mythology found in the *Kojiki* (712 A.D.), the *Nihongi* (720 A.D.), and in many texts written throughout the mediaeval period (from the twelfth to sixteenth centuries) as well as during the pre-modern period, Japan proposes a particular vision of its land which deserves thorough analysis.

The mythology does not provide any insight into a cosmogony, except to say that a series of formless divinities manifested themselves spontaneously, and that the last couple of such divinities had a form, and were a male (Izanagi) and a female (Izanami). These stood on a Heavenly Bridge and, thrusting into the ocean below a jewel-headed spear, churned the water. As they withdrew the spear, some drops of brine fell onto the surface of the ocean, where they coagulated to become an island. Descending to that island, the divine couple then, upon invitation on the part of the female divinity, engaged in sexual union. The result of this first union was a leech, which the couple did not find aesthetically satisfying, and therefore threw down "the cosmic drain" in a float made of reeds. Returning to Heaven to enquire of the reason of this unaesthetic production, they were told that the female should not invite, but the male. They decided to descend again to the island and, after the male had invited, they resumed their interaction, out of which the "Eight Islands" (Japan) were born. Continuing, the female deity gave birth to natural elements, such as the seas, the rivers, the mountains and the trees, water, and fire. Upon giving birth to fire, the female was so burnt that she passed away, although, in the last thrusts of her energy, she gave birth in her urine, her faeces, and her vomit, to divinities representing agriculture and sericulture.

Then the male deity, angered at fire, "killed" it. The description of the ritual killing of fire by his father is in the form of the description of the process of swordmaking.

After having killed his son, the male then followed his spouse to a realm under the earth, and asked her to return in order to continue the process of creation they were engaged in. She agreed, under the condition that she not be seen by him. Curious, he made some light and saw his spouse in the process of decay; horrified, he escaped, followed closely by demons and by his angered spouse. At the entrance of the netherworld, he placed a boulder, thus separating the realm of death and that of life. He then pronounced the words of repudiation from his spouse, who vowed to kill a number of his children everyday, upon which he vowed to give birth to even more. Thereafter, having been in contact with decay, he decided to purify himself in water, at the mouth of a river. Purifying his left eye, he gave birth to the solar divinity, emblem of the imperial lineage of Japan. Purifying his right eye, he gave birth to the lunar divinity, emblem of agricultural production. Purifying his nose, he gave birth to a violent divinity of which it could be said that it represents the industrial and military complex. Purifying the rest of his body, he gave birth to the divinities representing sea-faring.

The mythology then takes us into other realms leading to the smooth transition into history itself. One has therefore, the impression that the world of myth and that of time-in-history are not separated by much.

What is interesting from the particular vantage point we have chosen for this article is to notice that fire holds a mediating position between the production of the natural world through sexual interaction and the production of emblems of socio-cosmic character through a process of purification. Therefore, the world of nature is seen by ancient Japanese as preceding the "birth-and-death" of fire, whereas it could be said that whatever happens after the "death" of fire represents the world of culture. In this case, the death of fire ought to be understood to mean its control, which marked the death of fire in the realm of nature and its entrance into the realm of culture. That is why its "killing" is ritual and is expressed in terms in which one recognizes the production of swords, which to the ancient Japanese were the emblem of power and pacification or control. A further remark which may be made in order to demonstrate that Japanese mythology is indeed composed in a structural manner is that all the divinities of nature are born from the lower orifices of the feminine deity, whereas all the divinities related to culture (social structure, control over the seas surrounding the islands, grounding of legitimacy) are born from the head of the male divinity. Sex stands on the side of nature, whereas culture is represented by the processes of purification which are by far the dominant characteristic of Japanese ritual behavior, and a characteristic of Japanese life in general.

We have now shown that Japanese mythology makes extremely clear distinctions between the world of nature and that of culture, and that this opposition is marked by the events surrounding the appearance and the controlling of fire. But it should also be pointed out that the processes of purification responsible for the appearance of culture are all taking place in natural surroundings, and it is there that one must look for the particularly Japanese dialectic between nature and culture. Let us notice at this point that fire holds a double characteristic, being violent on its nature-side (volcano) and potentially violent (sword) on its culture-side. The divinities representing the powers of fertility in agriculture are all born out of substances which are themselves lukewarm (appearing just after the birth of fire) and representative of processes of change and natural transformation: decay, digestion. Thus nature has, in Japanese mythology, an ambivalent character: though it looks beautiful, it is also the realm of change, decay, and putrefaction, to which is opposed the purification of culture. The feminine deity represents the rotten, whereas the male divinity represents the pure.

This "rotten" and repulsive characteristic of nature remained in the perceptions of the Japanese for centuries, but it was also forgotten sometimes in the favor of a view advocating the beauty and the purity of nature. This is especially the case after centuries of processes of purification of nature at the hands of the cultivated people who invented gardens and flower arrangement, and the other great arts by which Japan likes to identify itself on the international cultural scene. It might be said that what has been termed "the Japanese love of nature" is actually the "Japanese love of cultural transformations and purification of a world which, if left alone, simply decays." So that the love of culture takes in Japan the form of a love of nature. It may be said that traditional interpreters of Japanese culture have failed to see this point, blinded as they were, perhaps, by a Western romanticism which is out of place.

Consequently, it is best, when attempting to investigate the position of nature and culture in the Japanese tradition, to attempt to shed our own cultural biases and tendencies.

In order to do this, perhaps a good method resides in systematically asking about the character of perception as it is proposed by the culture itself. But Japan, outside of the various Buddhist schools, does not offer philosophical treatises dealing with this issue in the way in which we are accustomed to. Instead, it offers a formidable wealth of materials from which we have to infer the particular phenomenology of perception which might be at work. In concrete terms: we can infer the Japanese philosophy of perception from poems, dramas, works of art and the like. And, it so happens that most of these seem to be representations of particular relationships to nature. Furthermore, one may, in order to investigate the relationship of the Japanese to nature, decide to research the various cults which are blatantly nature-related: mountain cults, waterfall cults, fire cults, sea cults, star cults, food cults, animal cults, and the like. This is a rather large order which cannot be realized in the scope of the present article, but we shall suggest from now on, the type of work which might be done in this direction in order to inform the expression of a particular ecophilosophy grounded in Japanese culture.

THE STUDY OF JAPANESE MINDSCAPES

I call *mindscapes* the various geographical areas chosen by the Japanese in the course of their history to either project onto them particular mental structures or representations of reality, or to infer from them particular representations helping them to establish meaning in experience. In either case we confront a dialectic between nature and culture, be it that nature is to be "decoded" in order to reveal its hidden meanings which are necessary in order to survive, or that culture is seen as the sum of actions which are informed by particular perceptions of the "being" of nature. These mindscapes are generally located in landscapes of great natural beauty which have been protected over the centuries and many of which form today the National Parks of Japan. It seems that the origin of these mindscapes is in ritual, when sacred space was defined in order to perform the rites of purification necessary to come into contact with the divinities which should guide human action. Natural elements form the basis of sacred space: a stone or a pillar, or a tree. These were in principle situated near sources of water, at the foot of mountains or at their top. It is there that specialists of ritual would manipulate fire and water, and play musical instruments in order to be possessed by the divine. This possession led to the uttering of "meaningless" sounds, which then had to be decoded and interpreted by specialists. In other words, it was thought that nature spoke a language which needed to be decoded. This aspect remained true for centuries in Japan, even in the texts proposing highest reflections on the philosophy of Buddhism. Some sacred spaces thus came to be seen as the natural abode of the divinities, and were not to be entered except at the time of ritual feasts; some of these, after the introduction of Buddhism, were forbidden to women, over whom the rotten characteristic of nature had been projected by mythology. Being the natural abode of the divinities, these areas became the focus of particular attention, and, as sedentary communities emerged, shrines were built for rituals. When Buddhism came in, temples were built next to the shrines, and associations between the Shintō divinities worshiped in the shrines and the Buddhist divinities worshiped in the temples were established, leading to complex combinatory systems of syncretism. The divinities worshiped in the shrines were often spirits of natural elements, or protectors of the community, or ancestral divinities; in most cases these merged to form a single complex deity. But the rituals show that we are always facing an attempt, on the part of the ritualists, to manipulate or influence nature with culture: food offerings are made, magical formulas which are believed to be the language of nature are expressed, thus making communication with nature possible. In the case of mountains, the entire area was seen as sacred; sometimes, the mountain itself was considered to be the "body" of the deity itself. Most often Buddhist temples were built next to these shrines, but as time passed, they came to be erected on the mountain itself: the ultimate in terms of culture could be realized in the deepest, or highest, or most ethereal parts of nature. The temples were granted tax-free domains at the foot of the mountains, thus gradually creating a large geographical unit that was under their spiritual influence and protection, in conjunction with the shrines. The result of these developments was the establishment of what could be called a sphere of influence, a self-contained unit of ritual and practice overlooking human activities in the plains. These cultic centers came to be the largest single land-owners of Japan during the mediaeval period. The sum of the symbols they expressed, of the rituals they performed, of the ideas they exuded or developed, forms the "mindscape." And in this mindscape, the presence of nature is overwhelming both in its outward appearance and in the culture it created; this is why a systematic study of these centers of nature/culture dialectic is needed. Their role in Japanese history is immense, for they regulated patterns of land-ownership and use, and gained large economic and political power. It can be advanced that the official separation between Shintō and Buddhism which was ordered by the government in 1868 had as one of its goals the fundamental change of land-ownership systems in Japan; therefore, the relationship between the cultic centers and the people living on that land became ever more tenuous, to the point of disappearing in the time-period of only one century. As a consequence, the contents of the relationship of people to nature changed drastically and followed other patterns of use that are not informed anymore by what goes on in the religious centers. This rearrangement of Japan may have cut the umbilical cord to ritual allowing people to deal with nature in a totally different way, which may have been what we call today: ecological.

AN EXAMPLE OF MINDSCAPE: KUNISAKI

Kunisaki is the name of a volcanic peninsula jutting out of the north-eastern corner of the Island of Kyushu. It is one of the most beautiful natural configurations of Japan, made up of a volcano which has a double peak and therefore receives the name of Futagoyama: "The Twin Mount." Its gentle slopes are covered with extraordinary rock formations and a luxuriant vegetation changing with the four seasons in a well-patterned rhythm. But they are also covered with about fifty Buddhist temples and many more Shintō shrines, which have been erected during the classical period of Japanese history; from the ninth century on. The reason why this cultic center of vast proportions became so important is that it was located near a most important Shintō shrine dedicated to the deity Hachiman, but also for a reason which should be of direct interest to us. This mountain came to be seen as the "natural form" of the most important scripture of Buddhism: the Lotus Sutra. Mountain practitioners of the classical period who specialized in this scripture noticed that there were as many natural valleys on the volcanic cone as there are chapters in the scripture: twenty-eight. They thus decided that each valley corresponded naturally to a chapter, or they came to the conclusion that each valley was the natural manifestation of a particular message. In order to realize this correspondance between "things" (nature) and "words" (culture), they then sculpted as many stones into the form of various Buddhas and Bodhisattvas as there are ideograms in the scripture in question: some sixty-thousand and several hundred statues. They placed these statues representing a word of the scripture on the paths they used in their ritual peregrinations while meditating on the meanings of the scriptures and on the meaning of their doing so while walking on the mountain. Very soon, it becomes difficult to make the following distinction: was it that nature spoke, or that culture was expressing the essence of nature? One could propose that the text had been "en-mountained" and that the mountain had been "textualized," and that nature thus spoke a most cultural discourse while culture expressed a basic harmony of meaning with that originally perceived to be in nature. This event in the evolution of religious practices in Japan could be seen as the manifestation of a philosophy of immanence, since that which had always been thought to be transcendant: the divine, was not to be "seen" in nature. Spinoza, the great Western philosopher of immanence, wrote that "a true idea must be in accordance with the object of which it is an idea" (*Ethics*, Vol. II, Axiom No. 6, Prop. No. 34 & 43). It could be said that the mindscape of Kunisaki expressed the same axiom in its own way.

After many vicissitudes of an economic and political nature, the Kunisaki Peninsula lost most of its economic basis allowing it to survive at this level, and gradually fell into disrepair. The separation of Buddhism and Shintō cooked up by the nineteenth century ideologues gave it a powerful blow. The afterwar period finished it. It is now the least populated area of Japan, a Quasi-National park of extreme beauty and poverty. The only industry is oranges and mushrooms. That is why the Japanese government is planning to build there its latest nuclear energy installation. In the meantime, many of the elegant statues representing the words of the Scripture of the Lotus have found their way into museums and "private collections." Nature and culture have been estranged, and I would submit that we suffer of this estrangement in every aspect of our being.

A ZEN EXAMPLE OF NATURE AND CULTURE

Zen has often been invoked in the West as the school of thought and practice which best represents the essence of Japanese culture, and it has been readily adopted by many Westerners as something akin to the ecological dream of contemporary marginals. While there is considerable ground to disagree with the statement that Zen is close to the "essence" of Japanese culture, or with the other statement which sees Zen as akin to some, we will not enter this discussion, and focus instead on the work of one of the greatest Zen masters of the mediaeval period: Dogen, for we believe that what he had to say about nature and culture in order to determine the ontological status of both is of direct relevance to our investigation.

In his main work entitled Shōbōgenzō, Dōgen proposes several chapters dealing exclusively with these questions. Of these we will retain two. The first one is entitled "Buddhist Scriptures," and questions the definition which was generally accepted by his students at the time. For most, the term "Buddhist Scriptures" referred to the written documents purported to be the sermons of the Buddha, and kept in the temple libraries: they had to be arduously studied if one wanted to progress on the path toward Awakening. Therefore, many doct students learned Chinese and read them with great care, and believed that the orthodoxy was to be found therein. But Dōgen, keeping the spirit of the Zen tradition as he had come to know it in China, disagreed with this position and proposed the following:

> *"In the Buddhist Scriptures you will find the Law which was taught by the Bodhisattvas, the Law which has been expounded by the various Buddhas. These scriptures are the*

tool of the Great Way. [. .] A Master necessarily and always understands clearly the meaning of the scriptures. To understand these scriptures is to make of them one's nation; it is to make of the scriptures one's own body and mind; it is to use them for someone else; it is to make of them one's own bed and walking. It is to make of them one's own father and mother, one's children and grandchildren. When I say that one should practice according to the scriptures, it means: according to a complete and clear understanding of what these scriptures are about. Thus, the behaviour of a master, like his washing his face or his drinking tea, is not different from the teachings to be found in scriptures; in fact, it is an old scripture itself. It happens that scriptures create masters: thus Ōbaku found his successor by beating Rinzai sixty times. A master was created when Reiun Shigon saw peach blossoms, or when Kyōgen Shikan heard a stone hit a bamboo, or when Śākyamuni saw Venus in the sky. All these are cases in which 'scriptures' have created masters. But it also happens that someone understands the real meaning of a scripture by opening his heart-mind's eye, and that he obtains that opening of the eye through the understanding of the real meaning of a scripture. What I call scripture, in fact, is the whole Universe understood at once in all ten directions. There is simply no time nor space that is not a 'scripture.' There are all kinds of scriptures: some expound a sublime truth, others express a trivial reason. Some are written with the words of another world; others are written with the words of animals, or written with the words of the enemies of the Buddhist Law, or with the words of human beings. Some indeed are written with the words of blades of grass or written with all kinds of trees.

Thus, the large, the short, the square, the round, the green, the yellow, the red, the white, all mixed through the vast universe in all directions, are the words of the 'scripture,' they are the appearance of the text we should be dealing with. This *is the Tool of the Great Way;* this *is what Buddhists should consider when they speak of 'scriptures.' "*

It could be said that Dogen proposed a certain type of phenomenology of existence according to which, people should study their mode of being and find out whether it was informed by an adequate perception of the being of nature. This master was not proposing that knowledge was to be found in texts, but rather in an understanding of modes of being which may be expressed in one way or another in some Buddhist scriptures. However, sole reliance on these written words was unacceptable to him, as I suppose that sole reliance upon a direct or immediate perception of nature and experience in it would not have been acceptable either. What is acceptable is a systematic, perhaps continuous examination of nature from the point of view of culture and of culture from the point of view of nature, thus leading to the establishment of some "ethics" or philosophy of action which would lead to a proper mode of being in the world, or to what Buddhists call Awakening and traceless enlightenment. The notion that the natural world was the "Book of God" is not purely Japanese; one finds it in China and in mediaeval Europe as well (one may refer to A.M. Crutius' work on mediaeval Europe), and in the United States, especially in the domain of painting and in the history of the establishment of the National Parks (a good work on painting from this perspective is Barbara Novak: *Nature and Culture in American Landscape Painting*). But the consciousness which is not regularly put in touch with that type of idea may not find the idea by itself, or even find it relevant. That is why Dōgen insisted on training and not on total immersion in nature only. Actually, Dōgen went beyond the notion that nature was the "Book of God," and proposed that the sounds of streams and the shape of mountains were the voice of the Buddha and the body of the Buddha themselves. But there is a slight difference which is enormously important: if the sound of a river is the voice of the Buddha expounding some profound truth about the nature of reality, how come that it cannot be immediately decodable? And if the shape of the mountains is the body of the Buddha and is adorned by as many splendours, how come we do not see the Buddha when trekking in our backyard? Should one get some training in a temple after learning classical Chinese before being able to see the Buddha in a tree? Is there then such a thing as "metaphysical jogging" or transcendental perception? Dogen would most probably say that the case of a doct and respectable student of dusty manuscripts taken from the caverns of northwestern China is as hopeless as the case of a jogger who will go to the bottom of his energy in a quest for vision. Another chapter of the Shōbōgenzō gives us some clues: it is entitled "The Voice of Streams and the Form of Mountains," of which I propose here some excerpts:

> *"In the country of the Great Song, there lived a man called Dong-bo. He was a true dragon in the ocean of calligraphy, and he studied the ocean of Buddhism under dragons and elephants. Able to swim in the greatest depths, he could also reach the highest clouds. When he was staying at Mount Lu, he achieved awakening while listening to the nocturnal babbling of a mountain stream. He then wrote a stanza, which he offered to the Master of Contemplation Chang-zong, and which reads:*

"The voice of streams is the sermon of the Buddha,
The form of mountains is His pure body;
Tonight I heard all teachings,
But how could I repeat them?"

Zong approved the stanza. [...] Is this example of Awakening while hearing a stream of some benefit to us, late comers? Is not deplorable that—how many times—we have let leak away the possibility of being converted by the sermon of the natural body of the Buddha? But then, should we not investigate how we listen to streams, how we look at the form of mountains? Should we hear them as being the absolute expression of truth in one verse, or is it only half of it, or is it all of it in eighty-four thousand stanzas? How regrettable that the voice and the body of the Buddha are hidden in mountains and streams! How rejoicing though, to know that there exists the possibility of a time for their manifestation! The voice of the Buddha never flinches, his body is never subject to appearance or to disappearance. Should we say that when we see it, it is close to us, that it is still close to us when we fail to see it? Is it appearing completely in the natural world, or is it showing only part of itself? Small indeed is the difference between all those years spent without seeing the Buddha in mountains and without hearing the Buddha in streams, and this particular night when they were . . .

Now, Bodhisattvas who study the Way have to start from the point of view that mountains flow and that streams do not move. The eve of the night when the poet apprehended the truth, he had gone to his master to ask some questions about the capacity of inanimate beings to expound the truth. Although he had not fully understood his master while listening to him, when the next night the sound of the stream was audible, it became one with the words of the master and great waves suddenly surged high up 'til they clashed against heaven. However the question is: what did shock the poet into Awakening? Was it the voice of the stream? Or was it the flux of words poured over him by his master? I harbor the idea that the talk of his master was still echoing in his mind and mixed in an obscure manner with the sound of the stream. Who then would want to debate whether the natural world gave only one cup or whether it gave a whole ocean? In other words: was it the poet who apprehended the truth? Or was it the natural world? Who will be clear-sighted enough to quickly see without doubt the sermon of the Buddha and his pure body?"

Of the many comments which could be made about this text we will retain only a few, to insist on the possibility that *if* nature can be perceived to be in some way the body of the Buddha, and *if* natural sounds can be perceived to be the expression of the Buddhist Law, then nature represents the ultimate, the absolute, the transcendant. This would have been self-evident to the Japanese of the time, who on the one hand spent time reflecting on the concept that transmigration is nirvana and on the other hand spent time writing poetry on the changes of seasons and saw in these changes (transmigration) the ultimate mode of being (nirvana). But the position espoused by Dōgen seems to be more complicated than pantheism; for, according to him, nature does not present itself to our ordinary perception as the ultimate. Only a process of purification of the sense organs, accompanied by meditation, allows one to reach the vision of the "Universe at Once in the Ten Directions." One recognizes here a fundamental aspect of Buddhist practice: rites of penitence are usually rites of purification of the sense organs which allow an adequate perception of the world. This perception, when achieved, is always qualified as "aesthetically pleasing." But ordinary perception on the part of people who are not trained leads astray and gives birth to a mode of being characterized by suffering, desire, loss. There is then a basic opposition between ordinary, "illusory" perception, and the "pleasurable sight" proposed by rites of penitence and by Dōgen; that is why he recommends that we should start by thinking that mountains walk and streams stay immobile. Pantheism would not advocate a systematic denial of ordinary perception.

Dōgen was not the only thinker of classical Japan to speak in such terms; Kūkai (773–835) before him, proposed extremely close perspectives on nature, and many a poet and thinker echoed these positions for centuries. Many Nō dramas represent these views as well. And in almost all cases, the relationship of nature to culture took the form of a "cosmic responsibility" on the part of man, who was seen as the agent of change in the natural realm. Dōgen wrote about this rather directly:

> *The ultimate is beyond the reach of the intellect of a beginner; the only thing to do, therefore, is to tread the way already shown by preceding saints. Searching for a master and for the Way, it may happen that you have to cross mountains and sail over oceans. Doing so, and asking masters, looking for good friends, you will see them come from Heaven like rain, and appear from the Earth like as many springs of water. When they guide you, they will have all animate beings preach, they will have all inanimate beings preach, and you will listen with your body, you will listen with your mind. To listen with the ears is of the domain of common experience, but to be able to listen with the eyes is the mode of being of*

absolute truth. When having visions of the Buddha, you will see it in yourself, in others; you will see it large, you will see it small. Do not be afraid of the large Buddha! Do not be disappointed by the small Buddha! You must see the Buddha as being the voice of streams and the form of mountains. In there are the Sermon of the Buddha and the Eighty-Thousand Stanzas, which are beyond communication and transcend all vision. Thus, the layman said: The higher, the harder," and the Buddha said: "Vast as Heavens and Oceans." In spring the pine-tree remains green, in autumn the chrysanthemum excels. That is all there is to it. [. . .] If you lose your confidence, it is necessary to concentrate on being straight and honest, and to repent to the former Buddhas. When doing so, the efficient virtue of this act will purify you. [. . .] When this pure confidence appears, the distinction between the self and others disappears, and its profit is extended to all sentient and non-sentient beings. Here is the content of the penitence: "I wish that, even if I have accumulated bad deeds in the past and have inborn causes of obstruction of the Way, I, relying on the Buddha's Way, realize it, and that all Buddhas and Patriarchs, by compassion, allow me to be liberated from the fetters of evil, annihilate in me all obstacles, and share with me this compassion allowing all merits to be shared by the vast universe. I wish that, by introspection, we all realize that the essence of the Buddha pervades the whole universe, that His Awakening is that of all beings, and that His perfection pervades more than the entire universe." (The Voice of Streams and the Form of Mountains.)

Why should practitioners of the Buddhist Way do penitence if they want to see the natural world as the Buddha? Perhaps because ethical behaviour is the only type of activity which allows nature to be fully perceived, and unethical behaviour, therefore, is the same as an improper attitude toward nature. If the natural world is the Buddha itself, then it is difficult to rape the earth, to open mountains, to sell their gold for profit and fame. Man is indeed the agent of change: when political figures behaved unethically, the result of their misdeeds always took the form of a natural cataclysm. Classical Japan developed a number of cults dedicated to members of society who had been wronged by politicians, who died in exile, and who came back as "angry spirits" (*onryō*) taking the form of tormentors of their enemies on the one hand, and the form of droughts, pestilence, earthquakes, on the other hand. These spirits had to be pacified, and these processes of pacification of cosmic disruption developed generally over a period of centuries. The most famous of such cults in classical Japan is that dedicated to Sugawara no Michizane (Kitano Tenjin), a scholarly political figure sent in exile by the Fujiwara House. He caused major disruptions in the natural cycles, and was the object of esoteric exorcism for a long time before he was completely pacified and seen as a benevolent deity patronizing the arts and scholarship. This cult is still popular today in Japan: many students flock to the shrines dedicated to that spirit before taking exams in universities, but it may be said that very few people know the origins or the exact content of the evolution of the cult as a major aspect of the dialectic between nature and culture. The same students worshiping the spirit of Michizane may very well become the next presidents of companies who have no interest whatsoever in ecology. The mindscape has been completely forgotten.

MANDALIZED MOUNTAINS

The Kji Peninsula is a rather large body of land extending south in the middle part of the main Island of Japan. It is today the site of two National Parks (Ise and Kumano) and one Quasi-National Park (Kōya-Ryūjin). The reason why the Peninsula is protected like this is not surprising now: it is the site of many temples and shrines, of mountain paths which have been for centuries trekked by mountain ascetics (*Yamabushi*) in their quest for the vision of that land as a metaphysical realm, a "Pure Land" on earth. Having presented in some detail the evolution of this area as one of Japan's foremost sacred spaces in another article (Grapard: "Flying Mountains and Walkers of Emptiness: toward a definition of sacred space in Japanese Religions." *History of Religions*, Chicago, 1982), I will summarize this phenomenon in the following lines. Esoteric Buddhism, the latest form of Buddhism developed in India, spread to Japan in the early ninth century, and had on Japanese culture a formidable impact. A highly ritual-oriented system of practices and a highly developed philosophical system, it is not easy to approach or to present globally. But let us propose here that it envisioned the universe from a double perspective which had to be unified through practice: the world of phenomenon and that of noumenon. These two worlds were symbolically represented in painting, as mandalas, which had to be penetrated ritually. The content of the practice was to approach the Triple Mystery of Mind, Speech and Body (that of the Buddha) ritually, this approach leading to one's fundamental capacities of the mind, of speech and of the physical body to be assimilated to those of the Buddha. In a word, it could be said that Esoteric Buddhism is a praxis of assimilation to higher levels of mode of being as defined by the tradition, or, perhaps better, an alchemic praxis purporting to transmute the mind of humans into that of the Buddha, during a single lifetime. From the point of view of Awakening, the world of the Buddha and that of humans is in a relationship of non-twoness. From the point of view of illusion, these worlds are set apart by incalculable realms. Therefore, a mandala is, from the point of view of Awakening, a pictographic representation of the world. It is, from the point of view of illusion, a representation of a transcendental realm. The goal of the praxis is to realize the view that both positions are neither true nor false, in virtue of the philosophical axiom that "illusion is awakening." We had mentioned earlier that ancient Japanese viewed mountains as residences for their divinities. What Esoteric Buddhism did in the Kii Peninsula was to imagine a huge mandala in the sky as if it were a diapositive, to use the sun as the projector, and to therefore have a projection of the mandala onto the mountainous area, in which each peak would be seen as the residence of a particular divinity of the mandala. As practitioners would, in a temple, sit in front of a painted mandala and enter it ritually, in this case of "natural mandala," they would pass through a gate in a shrine at the foot of mountains, and enter the imagined mountain mandala, perform on each natural peak the rites they would perform for each divinity in the painted mandala, and realize awakening at the summit of a particular mount located between the two major mandalized areas. The whole peninsula had therefore become a symbol of the realm of the Buddha, it was entirely sacralized by such practices. This is the deep reason why this area was protected by laws making it a National Park. But today the actual paths of the mountain ascetics are lost, no one knows the exact course, and the mountains are trekked by twentieth-century mountaineers escaping the noise of large cities. It is hard to know how much of the feelings responsible for the mandalization of such areas is re-created by these sunday trekkers.

THE MODERN CRISIS

In spite of houses made of natural elements and surrounded by gardens of great power to suggest the dialectic between nature and culture, in spite of a vast body of poetry and literature in which one can retrace the evolution of Japanese culture in its relationship to nature, and in spite of formidable religious systems which have addressed the question for a very long time, Japan turned the century and opened to the West in a catastrophic manner: rejecting much of its past, it emulated the power discourse of Europe and assimilated in no time the idea that nature is something to be controlled, rather than man's activities in it. In the process of "modernization" characterized by some as a success, in the process of development framed by the subsersive notion of relevance, and in the process of accumulating a wealth in total lack of balance with its natural boundaries, Japan has become a land destroyed and polluted. The only redeemable aspect of some of the policies taken by the government to curb pollution was the decision to translate the term "environmental pollution" by the term *kōgai,* which means, literally, "public harm." Japan has become well-known for its smog, for the Minamata disease, and for ailments caused by pollution. Little is done in education to assist technological attempts at reducing pollution with a systematic appreciation of the past, in which the Japanese could rediscover some fundamental attitudes which could become the basis for developing an environmental ethic, grounded in philosophical systems which had guided Japan for centuries and were forgotten in the aftermath of the war. But is it arrogant to suggest, as an outsider, that Japan has in its own cultural heritage, everything it takes to develop a new ethic for the modern age? And to suggest that this new ethic is not purely Japanese in a narrow sense, but is universally debatable and applicable? That in this heritage one discovers a fundamental universality of modes of being which are still "relevant" to the contemporary situation, East and West? But would education have the courage to be subsersive in the face of multinational corporations whose basic tenet is profit and whose cosmic responsibility in the face of "public harm" is the least of worries?

It is, after all, a matter of ethics.

Identification as a Source of Deep Ecological Attitudes

Arne Naess

Arne Naess is Professor of Philosophy at Oslo University. In his own words, he has written "too many books." These include Interpretations and Preciseness *(1953);* Ideology and Objectivity *(1956);* Four Modern Philosophers: Carnago, Wittgenstein, Heidegger, Sartre *(1965);* Communications & Argument *(1966);* Skepticism *(1968); and* Gandi & Group Conflict *(1974). Dr. Naess is a lover of mountains, deserts and sea and is a leader in the growing field of deep ecology.*

THE SHALLOW AND THE DEEP ECOLOGICAL MOVEMENT

In the 1960s two convergent trends made headway: a deep ecological concern, and a concern for saving deep cultural diversity. These may be put under the general heading "deep ecology" if we view human ecology as a genuine part of general ecology. For each species of living beings there is a corresponding ecology. In what follows I adopt this terminology which I introduced in 1973 (Naess 1973).

The term *deep* is supposed to suggest explication of fundamental presuppositions of valuation as well as of facts and hypotheses. Deep ecology, therefore, transcends the limit of any particular science of today, including systems theory and scientific ecology. *Deepness of normative and descriptive premises questioned* characterize the movement.

The difference between the shallow and deep ecological movement may perhaps be illustrated by contrasting typical slogans, here formulated very roughly:[1]

Shallow Ecology	Deep Ecology
Natural diversity is valuable as a resource for us.	Natural diversity has its own (intrinsic) value.
It is nonsense to talk about value except as value for mankind.	Equating value with value for humans reveals a racial prejudice.
Plant species should be saved because of their value as genetic reserves for human agriculture and medicine.	Plant species should be saved because of their intrinsic value.
Pollution should be decreased if it threatens economic growth.	Decrease of pollution has priority over economic growth.
Third World population growth threatens ecological equilibrium.	World population at the present level threatens ecosystems but the population and behavior of industrial states more than that of any others. Human population is today excessive.
"Resource" means resource for humans.	"Resource" means resource for living beings.
People will not tolerate a broad decrease in their standard of living.	People should not tolerate a broad decrease in the quality of life but in the standard of living in overdeveloped countries.
Nature is cruel and necessarily so.	Man is cruel but not necessarily so.

Deep ecological argumentation questions both the left-hand and the right-hand slogans. But tentative conclusions are in terms of the latter.

The shallow ecological argument carries today much heavier weight in political life than the deep. It is therefore often necessary for tactical reasons to hide our deeper attitudes and argue strictly homocentrically. This colors the indispensible publication, *World Conservation Strategy*.[2]

As an academic philosopher raised within analytic traditions it has been natural for me to pose the questions: How can departments of philosophy, our establishment of professionals, be made interested in the matter? What are the philosophical problems explicitly and implicitly raised or answered in the deep ecological movement? Can they be formulated so as to be of academic interest?

My answer is that the movement is rich in philosophical implications. There has however, been only moderately eager response in philosophical institutions.

The deep ecological movement is furthered by people and groups with much in common. Roughly speaking, what they have in common concerns ways of experiencing nature and diversity of cultures. Furthermore, many share priorities of life style, such as those of "voluntary simplicity." They wish to live 'lightly' in nature. There are of course differences, but until now the conflicts of philosophically relevant opinion and of recommended policies have, to a surprisingly small degree, disturbed the growth of the movement.

In what follows I introduce some sections of a philosophy inspired by the deep ecological movement. Some people in the movement feel at home with that philosophy or at least approximately such a philosophy, others feel that they, at one or more points, clearly have different value priorities, attitudes or opinions. To avoid unfruitful polemics, I call my philosophy "Ecosophy T," using the character *T* just to emphasize that other people in the movement would, if motivated to formulate their world view and general value priorities, arrive at different ecosophies: Ecosophy "A," "B," . . . , "T," . . ., "Z."

By an "ecosophy" I here mean a philosophy inspired by the deep ecological movement. The ending *-sophy* stresses that what we modestly try to realize is wisdom rather than science or information. A philosophy, as articulated wisdom, has to be a synthesis of theory and practice. It must not shun concrete policy recommendations but has to base them on fundamental priorities of value and basic views concerning the development of our societies.[3]

Which societies? The movement started in the richest industrial societies, and the words used by its academic supporters inevitably reflect the cultural provinciality of those societies. The way I am going to say things perhaps reflects a bias in favor of analytic philosophy intimately related to social science, including academic psychology. It shows itself in my acceptance in Ecosophy T of the theory of thinking in terms of "gestalts." But this provinciality and narrowness of training does not imply criticism of contributions in terms of trends or traditions of wisdom with which I am not at home, and it does not imply an underestimation of the immense value of what artists in many countries have contributed to the movement.

SELECTED ECOSOPHICAL TOPICS

The themes of Ecosophy T which will be introduced are the following:

The narrow self (ego) and the comprehensive Self (written with capital *S*)

Self-realization as the realization of the comprehensive Self, not the cultivation of the ego

The process of identification as the basic tool of widening the self and as a natural consequence of increased maturity

Strong identification with the whole of nature in its diversity and interdependence of parts as a source of active participation in the deep ecological movement

Identification as a source of belief in intrinsic values. The question of 'objective' validity.[4]

SELF-REALIZATION, YES, BUT WHICH SELF?

When asked about *where* their self, their "I," or their ego is, some people place it in the neighborhood of the *larynx*. When thinking, we can sometimes perceive movement in that area. Others find it near their eyes. Many tend to feel that their ego, somehow, is inside their body, or identical with the whole of it, or with its functioning. Some call their ego spiritual, or immaterial and not within space. This has interesting consequences. A Bedouin in Yemen would not have an ego nearer the equator than a whale-hunting eskimo. 'Nearer' implies space.

William James (1890: Chapter 10) offers an excellent introduction to the problems concerning the constitution and the limits of the self.

> *The Empirical Self of each of us is all that he is tempted to call by the name of me. But it is clear that between what a man calls* me *and what he simply calls* mine *the line is difficult to draw. We feel and act about certain things that are ours very much as we feel and act about ourselves. Our fame, our children, the work of our hands, may be as dear to us as our bodies are, and arouse the same feelings and the same acts of reprisal if attacked. And our bodies, themselves, are they simply ours, or are they us?*

The body is the innermost part of the material Self *in each of us; and certain parts of the body seem more intimately ours than the rest. The clothes come next. . . . Next, our immediate family is a part of ourselves. Our father and mother, our wife and babes, are bone of our bone and flesh of our flesh. When they die, a part of our very selves is gone. If they do anything wrong, it is our shame. If they are insulted, our anger flashes forth as readily as if we stood in their place. Our* home *comes next. Its scenes are part of our life; its aspects awaken the tenderest feelings of affection.*

One of his conclusions is of importance to the concepts of self-realization: "We see then that we are dealing with a fluctuating material. The same object being sometimes treated as a part of me, at other times is simply mine, and then again as if I had nothing to do with it all."

If the term *self-realization* is applied, it should be kept in mind that "I," "me," "ego," and "self" have shifting denotations. Nothing is evident and indisputable. Even *that* we are is debatable if we make the question dependent upon answering *what* we are.

One of the central terms in Indian philosophy is *ātman*. Until this century it was mostly translated with "spirit," but it is now generally recognized that "self" is more appropriate. It is a term with similar connotations and ambiguities as those of "self"—analyzed by William James and other Western philosophers and psychologists. Gandhi represented a *maha-ātman*, a *mahatma*, a great (and certainly very wide) self. As a term for a kind of metaphysical maximum self we find *ātman* in *The Bhagavadgita*.

Verse 29 of Chapter 6 is characteristic of the truly great *ātman*. The Sanskrit of this verse is not overwhelmingly difficult and deserves quotation ahead of translations.

sarvabhūtastham ātmānam
sarvabhutāni cā'tmani
Itsate yogayuktātmā
sarvatra samadarśanah

Radhakrisnan: "He whose self is harmonized by yoga seeth the Self abiding in all beings and all beings in Self; everywhere he sees the same."

Eliot Deutsch: "He whose self is disciplined by yoga sees the Self abiding in all beings and all beings in the Self; he sees the same in all beings."

Juan Mascaró: "He sees himself in the heart of all beings and he sees all beings in his heart. This is the vision of the Yogi of harmony, a vision which is ever one."

Gandhi: "The man equipped with *yoga* looks on all with an impartial eye, seeing *Atman* in all beings and all beings in *Atman*."

Self-realization in its absolute maximum is, as I see it, the mature experience of oneness in diversity as depicted in the above verse. The minimum is the self-realization by more or less consistent egotism – by the narrowest experience of what constitutes one's self and a maximum of alienation. As emperical beings we dwell somewhere in between, but increased maturity involves increase of the wideness of the self.

The self-realization maximum should not necessarily be conceived as a mystical or meditational state. "By meditation some perceive the Self in the self by the self; others by the path of knowledsge and still others by the path of works (*karma-yoga*) [*Gita*: Chapter 13, verse 24]. Gandhi was a *karma-yogi*, realizing himself through social and political action.

The terms *mystical union* and *mysticism* are avoided here for three reasons: First, strong mystical traditions stress the dissolution of individual selves into a nondiversified supreme whole. Both from cultural and ecological point of view diversity and individuality are essential. Second, there is a strong terminological trend within scientific communities to associate mysticism with vagueness and confusion.[5] Third, mystics tend to agree that mystical consciousness is rarely sustained under normal, everyday conditions. But strong, wide identification *can* color experience under such conditions.

Gandhi was only marginally concerned with 'nature.' In his *ashram* poisonous snakes were permitted to live inside and outside human dwellings. Anti-poison medicines were frowned upon. Gandhi insisted that trust awakens trust, and that snakes have the same right to live and blossom as the humans (Naess, 1974).

THE PROCESS OF IDENTIFICATION

How do we develop a wider self? What kind of process makes it possible? One way of answering these questions: There is a process of ever-widening ientification and ever-narrowing alienation which widens the self. The self is as comprehensive as the totality of our identifications. Or, more succinctly: Our Self is that with which we identify. The question then reads: How do we widen identifications?

Identification is a spontaneous, non-rational, but not irrational, process through which *the interest or interests of another being are reacted to as our own interest or interests*. The emotional tone of gratification or frustration is a consequence carried over from the other to oneself: joy elicits joy, sorrow sorrow. Intense identification obliterates the experience of a distinction between *ego* and *alter*, between me and the sufferer. But only momentarily or intermittently: If my fellow being tries to vomit, I do not, or at least not persistently, try to vomit. I recognize that we are different individuals.

The term *identification, in the sense used here*, is rather technical, but there are today scarcely any alternatives. 'Solidarity,' and a corresponding adjective in German, 'solidarisch,' and the corresponding words in Scandinavian languages are very common and useful. But genuine and spontaneous solidarity with others already presupposes a process of identification. Without identification, no solidarity. Thus, the latter term cannot quite replace the former.

The same holds true of empathy and sympathy. It is a necessary, but not sufficient condition of empathy and sympathy that one 'sees' or experiences something similar or identical with oneself.[6]

A high level of identification does not eliminate conflicts of interest: Our vital interests, if we are not plants, imply killing at least some other living beings. A culture of hunters, where identification with hunted animals reaches a remarkably high level, does not prohibit killing for food. But a great variety of ceremonies and rituals have the function to express the gravity of the alienating incident and restore the identification.

Identification with individuals, species, ecosystems and landscapes results in difficult problems of priority. What should be the relation of ecosystem ethics to other parts of general ethics?

There are no definite limits to the broadness and intensity of identification. Mammals and birds sometimes show remarkable, often rather touching, intraspecies and cross-species identification. Konrad Lorenz tells of how one of his bird friends tried to seduce him, trying to push him into its little home. This presupposes a deep identification between bird and man (but also an alarming mistake of size). In certain forms of mysticism, there is an experience of identification with every life form, using this term in a wide sense. Within the deep ecological movement, poetical and philosophical expressions of such experiences are not uncommon. In the shallow ecological movement, intense and wide identification is described and explained psychologically. In the deep movement this philosophy is at least taken seriously: reality consists of wholes which we cut down rather than of isolated items which we put together. In other words: there is not, strictly speaking, a primordial causal process of identification, but one of largely unconscious alienation which is overcome in experiences of identity. To some "environmental" philosophers such thoughts seem to be irrational, even "rubbish."[7] This is, as far as I can judge, due to a too narrow conception of irrationality.

The opposite of *identification* is *alienation*, if we use these ambiguous terms in one of their basic meanings.[8]

The alienated son does perhaps what is required of a son toward his parents, but as performance of moral duties and as a burden, not spontaneously, out of joy. If one loves and respects oneself, identification will be positive, and, in what follows, the term covers this case. Self-hatred or dislike of certain of one's traits induces hatred and dislike of the beings with which one identifies.

Identification is not limited to beings which can reciprocate: Any

animal, plant, mountain, ocean may induce such processes. In poetry this is articulated most impressively, but ordinary language testifies to its power as a universal human trait.

Through identification, higher level unit is experienced: from identifying with "one's nearest," higher unities are created through circles of friends, local communities, tribes, compatriots, races, humanity, life, and, ultimately, as articulated by religious and philosophic leaders, unity with the supreme whole, the "world" in a broader and deeper sense than the usual. I prefer a terminology such that the largest units are not said to comprise life *and* "the not living." One may broaden the sense of "living" so that any natural whole, however, large, is a living whole.

This way of thinking and feeling at its maximum corresponds to that of the enlightened, or yogi, who sees "the same," the *atman*, and who is not alienated from anything.

The process of identification is sometimes expressed in terms of loss of self and gain of Self through "self-less" action. Each new sort of identification corresponds to a widening of the self, and strengthens the urge to further widening, furthering Self-seeking. This urge is in the system of Spinoza called *conatus in suo esse perseverare,* striving to persevere in oneself or one's being (*in se, in suo esse*). It is not a mere urge to survive, but to increase the level of *acting out* (ex) *one's own nature or essence,* and is not different from the urge toward higher levels of "freedom" (*libertas*). Under favorable circumstances, this involves wide identification.

In western social science, self-realization is the term most often used for the competitive development of a person's talents and the pursuit of an individual's specific interests (Maslow and others). A conflict is foreseen between giving self-realization high priority and cultivation of social bonds, friends, family, nation, nature. Such unfortunate notions have narrow concepts of self as a point of departure. They go together with the egoism–altruism distinction. Altruism is, according to this, a moral quality developed through suppression of selfishness, through sacrifice of one's 'own' interests in favor of those of others. Thus, alienation is taken to be the normal state. Identification precludes sacrifice, but not devotion. The moral of self-sacrifice presupposes immaturity. Its relative importance is clear, in so far we all are more or less immature.

WIDENESS AND DEPTH OF IDENTIFICATION AS A CONSEQUENCE OF INCREASED MATURITY

Against the belief in fundamental ego–alter conflict, the psychology and philosophy of the (comprehensive) Self insist that the gradual maturing of a person *inevitably* widens and deepens the self through the process of identification. There is no need for altruism toward those with whom we identify. The pursuit of self-realization conceived as actualization and development of the Self takes care of what altruism is supposed to accomplish. Thus, the distinction egoism–altruism is transcended.

The notion of maturing has to do with getting out what is latent in the nature of a being. Some learning is presupposed, but thinking of present conditions of competition in industrial, economic growth societies, specialized learning may inhibit the process of maturing. A competitive cult of talents does not favor Self-realization. As a consequence of the imperfect conditions for maturing as persons, there is much pessimism or disbelief in relation to the widening of the Self, and more stress on developing altruism and moral pressure.

The conditions under which the self is widened are experienced as positive and are basically joyful. The constant exposure to life in the poorest countries through television and other media contributes to the spread of the voluntary simplicity movement (Elgin, 1981). But people laugh: What does it help the hungry that you renounce the luxuries of your own country? But identification makes the efforts of simplicity joyful and there is not feeling of moral compulsion. The widening of the self implies widening perspectives, deepening experiences, and reaching higher levels of activeness (in Spinoza's sense, not as just being busy). Joy and activeness make the appeal to Self-realization stronger than appeal to altruism. The state of alienation is not joyful, and is often connected with feelings of being threatened and narrowed. The "rights" of other living beings are felt to threaten our "own" interests.

The close connection between trends of alienation and putting duty and altruism as a highest value is exemplified in the philosophy of Kant. Acting morally, we should not abstain from maltreating animals because of their sufferings, but because of its bad effect on us. Animals were to Kant, essentially, so different from human beings, that he felt we should not have any moral obligations toward them. Their unnecessary sufferings are morally indifferent and norms of altruism do not apply in our relations to them. When we decide ethically to be kind to them, it should be because of the favorable effect of kindness on us — a strange doctrine.

Suffering is perhaps the most potent source of identification. Only special social conditions are able to make people inhibit their normal spontaneous reaction toward suffering. If we alleviate suffering because of a spontaneous urge to do so, Kant would be willing to call the act "beautiful," but not moral. And his greatest admiration was, as we all know, for stars and the moral imperative, not spontaneous goodness. The history of cruelty inflicted in the name of morals has convinced me that increase of identification might achieve what moralizing cannot: beautiful actions.

RELEVANCE OF THE ABOVE FOR DEEP ECOLOGY

This perhaps rather lengthy philosophical discourse serves as a preliminary for the understanding of two things: first, the powerful indignation of Rachel Carson and others who, with great courage and stubborn determination, challenged authorities in the early 1960s, and triggered the international ecological movement. Second, the radical shift (see Sahlins, 1972) toward more positive appreciation of nonindustrial cultures and minorities – also in the 1960s, and expressing itself in efforts to "save" such cultures and in a new social anthropology.

The second movement reflects identification with threatened cultures. Both reactions were made possible by doubt that the industrial societies are as uniquely progressive as they usually had been supposed to be. Former haughtiness gave way to humility or at least willingness to look for deep changes both socially and in relation to nature.

Ecological information about the intimate dependency of humanity upon decent behavior toward the natural environment offered a much needed rational and economic justification for processes of identification which many people already had more or less completed. Their relative high degree of identification with animals, plants, landscapes, were seen to correspond to *factual relations* between themselves and nature. "Not man apart" was transformed from a romantic norm to a statement of fact. The distinction between man and environment, as applied within the shallow ecological movement, was seen to be illusory. Your Self crosses the boundaries.

When it was made known that the penguins of the Antarctic might die out because of the effects of DDT upon the toughness of their eggs, there was a widespread, *spontaneous* reaction of indignation and sorrow. People who never see penguins and who would never think of such animals as "useful" in any way, insisted that they had a right to live and flourish, and that it was our obligation not to interfere. But we must admit that even the mere appearance of penguins makes intense identification easy.

Thus, ecology helped many to know more *about themselves*. We are living beings. Penguins are too. We are all expressions of life. The fateful dependencies and interrelations which were brought to light, thanks to ecologists, made it easier for people to admit and even to cultivate their deep concern for nature, and to express their latent hostility toward the excesses of the economic growth societies.

LIVING BEINGS HAVE INTRINSIC VALUE AND A RIGHT TO LIVE AND FLOURISH

How can these attitudes be talked about? What are the most helpful conceptualizations and slogans?

One important attitude might be thus expressed: 'Every living being has a *right* to live.' One way of answering the question is to insist upon the value in themselves, the autotelic value, of every living being. This opposes the notion that one may be justified in treating any living being as just a means to an end. It also generalizes the rightly famous dictum of Kant "never use a person solely as a means." Identification tells me: if *I* have a right to live, *you* have the same right.

Insofar as we consider ourselves and our family and friends to have an intrinsic value, the widening identification inevitably leads to the attribution of intrinsic value to others. The metaphysical maximum will then involve the attribution of intrinsic value to all living beings. The right to live is only a different way of expressing this evaluation.

THE END OF THE WHY'S

But why has *any* living being autotelic value? Faced with the ever returning question of "why?," we have to stop somewhere. Here is a place where we well might stop. We shall admit that the value in itself is something shown in intuition. We attribute intrinsic value to ourselves and our nearest, and the validity of further identification can be contested, and *is* contested by many. The negation may, however, also be attacked through series of "whys?" Ultimately, we are in the same human predicament of having to start somewhere, at least for the moment. We must stop somewhere and treat where we then stand as a foundation.

The use of "Every living being has a value in itself" as a fundamental norm or principle does not rule out other fundamentals. On the contrary, the normal situation will be one in which several, in part conflicting, fundamental norms are relevant. And some consequences of fundamental norms *seem* compatible, but in fact are not.

The designation "fundamental" does not need to mean more than "not based on something deeper," which in practice often is indistinguishable from "not derived logically from deeper premises." It must be considered a rare case, if somebody is able to stick to one and only one fundamental norm.(I have made an attempt to work with *a model* with only one, Self-realization, in Ecosophy T.)

THE RIGHT TO LIVE IS ONE AND THE SAME, BUT VITAL INTERESTS OF OUR NEAREST HAVE PRIORITY OF DEFENSE

Under symbiotic conditions, there are rules which manifest two important factors operating when interests are conflicting: vitalness and nearness. The more vital interest has priority over the less vital. The nearer has priority over the more remote – in space, time, culture, species. Nearness derives its priority from our special responsibilities, obligations and insights.

The terms used in these rules are of course vague and ambiguous. But even so, the rules point toward ways of thinking and acting which do not leave us quite helpless in the many inevitable conflicts of norms. The vast increase of consequences for life in general, which industrialization and the population explosion have brought about, necessitates new guidelines.

Examples: The use of threatened species for food or clothing (fur) may be more or less vital for certain poor, nonindustrial, human communities. For the less poor, such use is clearly ecologically irresponsible. Considering the fabulous possibilities open to the richest industrial societies, it is their responsibility to assist the poor communities in such a way that undue exploitation of threatened species, populations, and ecosystems is avoided.

It may be of vital interest to a family of poisonous snakes to remain in a small area where small children play, but it is also of vital interest to children and parents that there are no accidents. The priority rule of nearness makes it justifiable for the parents to remove the snakes. But the priority of vital interest of snakes is important when deciding where to establish the playgrounds.

The importance of nearness is, to a large degree, dependent upon vital interests of communities rather than individuals. The obligations within the family keep the family together, the obligations within a nation keep it from disintegration. But if the nonvital interests of a nation, or a species, conflict with the vital interests of another nation, or of other species, the rules give priority to the "alien nation" or "alien species."

How these conflicts may be straightened out is of course much too large a subject to be treated even cursorily in this connection. What is said only points toward the existence of rules of some help (For further discussion, see Naess [1979]).

INTRINSIC VALUES

The term "objectivism" may have undesirable associations, but value pronouncements within the deep ecological movement imply what in philosophy is often termed "value objectivism" as opposed to value subjectivism, for instance, "the emotive theory of value." At the time of Nietzsche there was in Europe a profound movement toward separation of value as a genuine aspect of reality, on a par with scientific, "factual" descriptions. Value tended to be conceived as something projected by man into a completely value-neutral reality. The *Tractatus Philosophico-Logicus* of the early Wittgenstein expresses a well-known variant of this attitude. It represents a unique trend of *alienation of value* if we compare this attitude with those of cultures other than our technological-industrial society.

The professional philosophical debate on value objectivism, which in different senses – according to different versions, posits positive and negative values independent of value for human subjects – is of course very intricate. Here I shall only point out some kinds of statements within the deep ecological movement which imply value objectivism in the sense of intrinsic value:

> Animals have value in themselves, not only as resources for humans.
>
> Animals have a right to live even if of no use to humans.
>
> We have no right to destroy the natural features of this planet.
>
> Nature does not belong to man.
>
> Nature is worth defending, whatever the fate of humans.
>
> A wilderness area has a value independent of whether humans have access to it.

In these statements, something *A* is said to have a value independent of whether *A* has a value for something else, *B*. The value of *A* must therefore be said to have a value inherent in *A*. *A* has *intrinsic value*. This does not imply that *A has* value for *B*. Thus *A* may have, and usually does have, both intrinsic and extrinsic value.

Subjectivistic arguments tend to take for granted that a subject is somehow implied. There "must be" somebody who performs the valuation process. For this subject, something may have value.

The burden of proof lies with the subjectivists insofar as naive attitudes lack the clear-cut separation of value from reality and the conception of value as something projected by man into reality or the neutral facts by a subject.

The most promising way of defending intrinsic values today is, in my view, to take gestalt thinking seriously. "Objects" will then be defined in terms of gestalts, rather than in terms of heaps of things with external relations and dominated by forces. This undermines the subject-object dualism essential for value subjectivism.

OUTLOOK FOR THE FUTURE

What is the outlook for growth of ecological, relevant identification and of policies in harmony with a high level of identification?

A major nuclear war will involve a setback of tremendous dimensions. Words need not be wasted in support of that conclusion. But continued militarization is a threat: It means further domination of technology and centralization.

Continued population growth makes benevolent policies still more difficult to pursue than they already are. Poor people in megacities do not have the opportunity to meet nature, and shortsighted policies which favor increasing the number of poor are destructive. Even a small population growth in rich nations is scarcely less destructive.

The economic policy of growth (as conceived today in the richest nations of all times) is increasingly destructive. It does not *prevent* growth of identification but makes it politically powerless. This reminds us of the possibility of significant *growth* of identification in the near future.

The increasing destruction plus increasing information about the destruction is apt to elicit strong feelings of sorrow, despair, desperate actions and tireless efforts to save what is left. With the forecast that more than a million species will die out before the year 2000 and most cultures be done away with, identification may grow rapidly among a minority.

At the present about 10% to 15% of the populace of some European countries are in favor of strong policies in harmony with the attitudes of identification. But this percentage may increase without major changes of policies. So far as I can see, the most probable course of events is continued devastation of conditions of life on this planet, combined with a powerless upsurge of surrow and lamentation.

What actually happens is often wildly "improbable," and perhaps the strong anthropocentric arguments and wise recommendations of *World Conservation Strategy* (1980) will, after all, make a significant effect.

NOTES

[1] For survey of the main themes of the shallow and the deep movement, see Naess (1973); elaborated in Naess (1981). See also the essay of G. Sessions in Schultz (1981) and Devall (1979). Some of the 15 views as formulated and listed by Devall would perhaps more adequately be described as part of "Ecosophy D" (D for Devall!) than as parts of a common deep ecology platform.

[2] Commissioned by The United Nations Environmental Programme (UNEP) which worked together with the World Wildlife Fund (WWF). Published 1980. Copies available through IUNC, 1196 Gland, Switzerland. In India: Department of Environment.

[3] This aim implies a synthesis of views developed in the different branches of philosophy – ontology, epistemology, logic, methodology, theory of value, ethics, philosophy of history, and politics. As a philosopher the deep ecologist is a "generalist."

[4] For comprehensive treatment of Ecosophy T, see Naess (1981, Chapter 7).

[5] See Passmore (1980). For a reasonable, unemotional approach to "mysticism," see Stahl (1975).

[6] For deeper study more distinctions have to be taken into account. See, for instance, Scheler (1954) and Mercer (1972).

[7] See, for instance, the chapter "Removing the Rubbish" in Passmore (1980).

[8] The diverse uses of the term *alienation (Entfremdung)* has an interesting and complicated history from the time of Rousseau. Rousseau himself offers

interesting observations of how social conditions through the process of alienation make *amour de soi* change into *amour propre*. I would say: How the process of maturing is hindered and self-love hardens into egotism instead of softening and widening into Self-realization.

REFERENCES

Elgin, Duane, 1981. *Voluntary Simplicity*. New York: William Morrow.

James, William. 1890. *The Principles of Psychology*. New York. Chapter 10: The Consciousness of Self.

Köhler, W. 1938. *The Place of Value in a World of Facts*. New York: On thinking in terms of gestalts.

Meeker, Joseph W. 1980. *The Comedy of Survival*. Los Angeles: Guild of Tutor's Press.

Mercer, Philip. 1972. *Sympathy and Ethics*. Oxford: The Clarendon Press. Discusses forms of identification.

Naess, A. 1973. *"The Shallow and the Deep, Long Range Ecology Movement," Inquiry* 16: (95–100).

– – –. 1974. *Gandhi and Group Conflict*. 1981, Oslo: Universitetsforlaget.

– – –. 1981. *Ekologi, samhälle och livsstil. Utkast til en ekosofi*. Stockholm: LTs förlag.

– – –. 1979. *"Self-realization in Mixed Communities of Humans, Bears, Sheep and Wolves," Inquiry*, Vol. 22, (pp. 231–241).

Passmore, John. 1980. *Man's Responsibility for Nature*. 2nd ed., London: Duckworth.

Rodman, John. 1980. "The Liberation of Nature," *Inquiry 20: (83–145).*

Sessions, George. 1981. Ecophilosophy (mimeo), Department of Philosophy, Sierra College, Rocklin, CA. Survey of literature expressing attitudes of deep ecology.

Sahlins, Marshall. 1972. *Stone Age Economics*. Chicago: Aldine.

Scheler, Max. 1954. *The Nature of Sympathy*. London: Routledge & Keegan, Paul.

Schultz, Robert C. and J. D. Hughes (eds.). 1981. *Ecological Consciousness*. University Press of America.

Schumacher, E. F. 1974. *The Age of Plenty: A Christian View*. Edinburgh: Saint Andrew Press.

Stahl, Frits. 1975. *Exploring Mysticism*. Berkeley: University of California Press.

Stone, Christopher D. 1974. *Should Trees Have Standing?* Los Altos, CA: Kaufmann.

World Conservation Strategy 1980. Prepared by the International Union for Conservation of Nature and Natural Resources (IUCN).

Survival or Transcendence — A Dialogue with Paolo Soleri

An Interview at the Cosanti Foundation Scottsdale, Arizona 1982

Soleri/Tobias/Gosney

Paolo Soleri was born in Turin, Italy and received his doctor of architecture degree in Italy. He has lived and worked in Arizona since 1956. Soleri is an architect, philosopher and craftsman. He has become known throughout the world for his innovative urban designs, called arcologies, and for his ceramic and metal windbells and sculpture.

Michael Tobias, formerly an Assitant Professor of Environmental Affairs and the Humanities at Dartmouth College, with a Ph.D. from the University of California-Santa Cruz.

Michael Gosney is a writer/artist with a diversified communications background. He is the publisher of Avant Books.

Tony Brown has been an architect in residence at Arcosanti for seven years.

Robert Radin is an entrepreneur with special expertise in global agriculture, textiles and low-income housing.

Ed Roxburgh is a California painter whose artwork is recognized internationally.

COMPLEXITY VS. COMPLICATION

Tobias: When someone hears the name Paolo Soleri and Arcosanti the first impressions are of utopia, socialism, the imposition of communal life as opposed to good old American hedonism. I want to know what Arcosanti is *asking* about America in terms of what really is ethical, what really is ecological. Right from the core. What do you do, how does that work? How are you going about it? What makes you feel this is the way of the future?

Soleri: If we start with American society (which we can equate very much with Western society, Europe, Canada, and so on), my first observation would be that we are mixing up the practical with the real. Unfortunately, the consequences are quite often not very positive nor very promising. I make the distinction between that which is real and important, and the immediate reward or immediate results sought by the practical mind. It's like pretending, or deceiving ourselves into believing that we have solved a medical problem by taking care of the aches. We segregate analysis and think *Eureka!* We've solved the problem. Or we can go from here to there in perfect comfort. I think that's what the practical bend in mind does very well and the auxiliary is technology which stands to be a limited, but effective solution to problems that are segregated so as to find solutions. So, by adding up all those small triumphs of the tales of reality we miss what reality might be which is not just a sum of those parts. It is more than the sum of parts. And I think the real requires a reconsideration of this capacity for analytical inquiries. The answers which are remaining within the context of analysis instead of looping back to synthesis are no answers at all since they are not capable of responding to the encompassing problems.

Tobias: What in your mind are some of the immediate clues to those large scale problems that are missed by the practical mind?

Soleri: Evidently, the ecological disaster that we have been moving into is a very real example of practical solutions that have tended to become the nemesis of that which we are dealing with. So as a farmer, I might do a beautiful job with my fifty acres of land and then discover chaos surrounding my little plot of ordered grounds. So it's the old story of genentropy being created at the expense of disorder. I have to know where I come from more and more. The more I deal with the future, the more I have to know what the past is. Otherwise, there's no way of doing anything in the future.

Tobias: This is what interests me. Okay, the very verbs that you're employing right now – *deal* and *know*. What do they mean?

Soleri: Deal is to proceed in implementing something which doesn't exist. It's just a wish or an anticipation or what you might call a plan. Knowing is a perception and a rationalization of something that has come about and now is stored somewhere. So in order for me to do any planning I need to know what is the reality which I'm going to manipulate in order to produce the plan, to implement the plan. If I don't know what the material I'm going to use is, my planning is even more abstract, even more arbitrary, so the knowledge – whatever knowledge one can have of one's own origin – I think is very important in order to anticipate.

Tobias: As you probe backwards do you identify any period in history when a model was put forth that intimated the kind of future harmony you are oriented towards in your work? Was there any period in medieval Europe, in the hunting and gathering societies, during the Renaissance?

Soleri: I wish I knew history so I could give you an answer. (laughter) But in a way any religious model contains at least some of the elements of the model I need in order to participate in making reality.

Tobias: The future.

Soleri: Yes.

Gosney: The ethical and moral models set forth by the religions, you mean?

Soleri: Yes, and also the performing models of the norns which are allowing for the performance of anything. The thing which I find very funny or not very funny is that I always have to take those models and look at them this way because they are upside down as far as I'm concerned. So I have to take them and turn them over and then they begin to make sense.

Tobias: How so?

Soleri: Well, because my notion is that if I'm presented with a reality which originally was absolute and perfect then I have a terrible problem to make any sense, to give any meaning, to talk in any sense about the legitimacy or the justness of anything. If the beginning was perfect what's all this fuss about? Let's have the beginning.

Gosney: Could that be some type of model for ultimate spiritual reality, original spiritual reality, totally apart from the physical world that we're going back towards . . . ?

Soleri: That's a dualism I won't have time to deal with. The notion of the *fall*, spirit that made matter and then matter not being up to spirit, seems to me very arbitrary. I can explain the wanting nature of matter because evidently we're made of matter and we find that we are faulty. I would say it's lucky that matter is, because only through matter, I can be, no matter how faulty I might be. Why not try and find out, attempt to do in such a way as to make matter less faulty. By making it into spirit.

Tobias: Or as Bertrand Russell put it, "What is mind? No matter and what is matter? Never mind."

Radin: Is it necessary to maintain the disorder or to let it destroy itself and hope that from those ashes something new surrounding the fifty acres would be engendered?

Soleri: I'm not endorsing the notion that I create order at the expense of whatever the cost might be. I believe that those fifty acres belong to the surrounding acreage. In general things are happening in positive ways if the degree of complexity of the total system is improved; things are going badly if the degree of complexity of the total system is diminished. Now what was the total system? A good question. I think probably we could limit the total system to this planet just for the moment. And then suggest that if the intervention of me, you, the trees and everything else, is such as to increase the complexity of what we call the biosphere or noosphere or whatever, then we are on the road to better things. If all those interventions are going to be diminishing the total complexity, then we are on the road to disaster.

Tobias: How do you define complexity?

Soleri: In a way, the degree of liveliness that results from the interaction of the participant, because I feel that complexity ultimately can be equated with what I might call liveliness or consciousness or sensitization and so on.

Radin: Do you infer, or do I hear you inferring that *all*things should be obliged to participate?

Soleri: No, but it is participating. Obliged or not, it is inescapably participatory. Because there is enough interdependence and dependence to make sure that the system is ultimately one system.

Brown: Do you think, Paolo, that there is no such thing as negative complexity?

Soleri: Yes, I tend to call that *complication* more than complexity.

Brown: But in a sense, complexity is a process that goes on and on; this may be a simplistic kind of example but modern technological complexity has brought us to the point of really polluting the biosphere, a form of complexity through human intervention in the material world. Do you see that as ultimately positive?

Soleri: I interpret that in a slightly different way, probably. I'm, let's say, the physiological and mental organism. And then I have a camera and all the equipment that I am working with. If the addition of this complexity (that is, if I graft this system to my own) and the end result is diminishing complexity, then I am in the presence of a negative process. It might well be that I am so addicted to the camera that I am not able to see without it. Well, at that point, I might be handicapped. My sensitivity and my ability to perceive and enjoy, to make something out of the environment, might be diminished.

Brown: Do you see the augmentation of individual complexity as necessarily essential for the augmenttion of societal complexity? For example, if we are all plugged into a computer terminal it might be very simplifying for the individual, but for society as a whole, very complexing. Or do you see the increase in individual complexity, spirituality and consciousness as absolutely essential for the increase of the whole society.

Soleri: I think the answer would be in exercising some kind of a very sharp analysis to find out what's happeninng and how it's happening. But I think ultimately that in order to have the greatest complexity, both parts and the totality have to be at the maximum capacity of complexity. One thing that technology clearly does is simplify, break down the complex and then come out with complicated devices which are very effective in a very narrow range.

Brown: So you would be in agreement with, say, the ecological movement that we should begin to look at the control of technology. We should begin to see technology as the major factor deteriorating the quality of life on the planet?

POLLUTION

Soleri: I would say that we are too nonchalant by always assuming that because we add a new technological tool we are going for greater complexity. We might well blind ourselves or take short cuts that, instead of enriching the system, impoverish it. When I'm told, for instance, that I shouldn't build a dam because this little fish might be sacrificed then I have to think twice. I understand the danger of saying well today we sacrifice the little fish, tomorrow the little frog, and so on. Sacrifice by sacrifice until we find out we're alone, the only life form, having gone into the ultimate sacrifice one step at a time. But that should not hide the fact that that little fish may ultimately find his niche: I may push him aside for the moment, then he may then go and do something else. Now that something else might be per se evil or might be per se ambiguous or might end up being very important.

Brown: So, you don't see non-intervention in the environment as a solution to the ecological crisis?

Soleri: No, because if there's an earthquake and a mountain collapses, making itself into a dam, with the result being your lake over there, I have a hard time seeing that as a positive ecological or divine providential act while my building a dam somewhere else, is an evil intrusion in the natural process. I put spirit where perhaps spirit is not. Which is in the earthquake coming at a certain spot at a certain moment and developing into a certain alteration. So on one side I have the pretension that whatever happens in nature is right and it is almost balanced by the other pretense that almost anything we do is wrong.

Radin: But nature has created the perfect environment, the result of which is man, the most perfect creature. Why try to alter that perfection?

Soleri: No, I don't buy that. We are what we are because there's been a sequence of processes which somehow one step at a time have defined man, but we could be a thousand times more perfect or a thousand times less perfect. I don't see any perfection in the processes of nature creating that.

Radin: But how did we gain this ascent? Why change the environment? Why not keep the environment which has taken you from the trees to Arizona?

Soleri: By that exact reasoning the tree doesn't have the right to be there. Because before that tree was there, there was an earlier tree which was much more primitive; and before that primitive tree was there, another creature which was even more primitive and all these creatures if they had a mind would say "hey, I'm in the image of god and nothing's going to change me because I'm in the image of god." So we de-escalate down to a moment when we find out there's only stone and water and fire. Even there we have to go back and we end up with hydrogen and the big bang and we say, "well, this is the real stuff, all the rest is intrusion — it's pollution."

Radin: But in that setting man proceeded up his own destiny. We certainly have progressed in the last million years in the right direction.

Soleri: Why then not accent that and say we are one of the steps in this ladder towards who knows what. If we are one intermediate step then evidently to say there is not only a mistake, it's a sin.

Radin: But it is also quite threatening to change the factory that produced the perfect machine.

Soleri: Well, no, there is not a perfect machine. That is the point I'm very much with.

Radin: Not perfect enough?

Soleri: No.

Gosney: We already have changed, changed the machine. Our existing cities have already changed us.

Radin: I've built things, I'm not pushing the big cities. I think if it went over to the other direction of becoming little hamlets and little villages that man might get to his next step in his chain more successfully. As you said, maybe with this high density, high technology, he's going to blow himself up. It certainly has indications.

Soleri: Yes, but the fact there exists a danger of pathology doesn't say we should remain what we have been.

Radin: It's a conflict. Three hundred years ago there wasn't the conflict of east and west, so we didn't create the hydrogen bomb. Man seems to create what he needs for the moment and it seems to me that the more density the more apprehension, the more fear, possibly the more destructive weaponry and so on.

Soleri: Yes, but again we can take it to extremes. We could say, Adam remarked to Eve, "We're the least densely populated form on the planet," is that the ideal? Or is it one billion? Who knows? To connect directly the positive with density? The capacity for love with density and so on.

POPULATION

Radin: But you seem to me, for the few moments we're here, to be sensitive, to be responsive, to be on a quest. Yet your setting is not Century City, your setting is not Chicago, your setting is here under your fig tree. A single man under a single tree.

Soleri: Well yes, but number one, what we're building there is slightly different from what we're building here. And what we're building here is not yet what we are seeking. So in a way my ideal is Chicago. It's not the fig tree and a chair under the fig tree. But I have reasons for believing that Chicago is closer to "god" than this chair under the fig tree. Not that the two are exclusive but there is definitely something that has to do with number, which has to do also with richness. And that's part of the complexity; what is complexity: more and more things using less and less media.

Gosney: That's a good general definition of complexity.

Tobias: That's basically the definition of stability in any ecosystem. A climax forest has reached that state of stability, it's very complex but it's of a higher order of stability than a preceding forest.

Soleri: Okay then, the next question would be, is this forest the icon that we are to contemplate forever? Many billions of years from now is the forest still a valid notion? Quite probably not. It wouldn't make any sense. It's terribly sensical now but we must watch so that we don't fossilize ourselves and the forest by saying there's nothing better than the forest. For the moment the forest is a tremendous asset and we should be very careful in getting rid of it, destroying it, not just because of our own sake, but for the sake of reality in general. But we have to be very careful because by saying that the forest is an eternal datum that is sacred and cannot be altered, we are speaking bigotry.

Tobias: Yet the very declaration which comes from our self-consciousness is prone towards a vanity that is unprecedented in evolution and that has its own risks.

Soleri: Sure.

Tobias: The forest never looks at itself and says, "I'm not going to be here in a million years, therefore I ought to work expeditiously to do something else.

Soleri: We know it's not going to be there anyhow.

Tobias: Well. We don't know that for sure, even examining the record of vegetation in the past, there has always been – since 400 million years ago – some kind of land cover. It doesn't conform to the same rudiments as today's cover . . . but . . .

Soleri: What I mean is that no matter what the past is telling us, that we know that in a certain span of time Earth is not going to be capable of supporting the forest because it's probably going to volatilize or be burned out or it's going to become a cold piece of stone, whatever. So it's quite clear that if there is in us a notion of fundamental justness, then the notion has come out of the model which is not the model of the eternal forest.

Tobias: But then how sensible is it to orient one's life to a model which won't actually be applicable for many millemiums? That's the other question.

THE ABSOLUTE

Soleri: Because we are after eternity, we are after the absolute. That's why we're inventing religions. Why be religious if you don't need those things? Forget it, let's be nice good pagans enjoying life, having a ball and forgetting about it. (laughter)

Tobias: And terribly polytheistic as pagans. And there was apparently in that free-for-all an even greater need of the absolute in that it was invested in every other object.

Soleri: Okay, then if we seriously talk about the absolute, we should be serious about it. And to be serious about it is to accept the notion that earth has a life span and beyond that lifespan there's no earth. But there might be something else far more incredible than the Earth, no matter how incredible the Earth is.

Brown: So within that allotted time span, presumably there is some process we have to reach and what do you see as that process?

Soleri: In very raw terms, that is the process of a reality that is becoming more and more complex, because by becoming more and more complex, it is become more and more able to read itself, to understand itself, to develop anticipatory models that are becoming more and more powerful in making prophecies, fulfillments and so on.

Brown: So the only hope for the Earth, as it were, is to get to the point where it can transcend the Earth itself once the Earth is destroyed.

Soleri: In absolute terms yes, I don't see any possibility of straying away from that. But if we want to be practical, then we say, okay, let's forget about religion and absolutes and so on. There still is a difference between the practicality of the Western man and the realism that Western man could develop, even staying away from the notion of absolute.

Brown: How does that absolute picture translate into present action, in terms of environmental questions?

Soleri: Again, in my simple-minded way, it's because I find this paradigm of complexity an absolute paradigm, so no matter what the span of time is that would interest us, if we want to have a richer life we have to have more complex interdependence and inter-relationships and so on. This will enable us to be more conscious of what surrounds us, what's inside of us and how tremendously difficult it is to coordinate those things. I tend to believe that there's a triggering point where quantity becomes quality.

Tobias: What do you base that belief on?

THE BRAIN

Soleri: Not that I know what the brain or the cortex is but I would say that there's no fundamental difference between the nervous system of a very primitive creature and my nervous system. The only difference is that the number of components which goes for the number of relationships again goes for the number of responses it escalates in such a marvelous way that at a certain point that very simple action and response whatever you call it, becomes thinking, becomes mind, and as it becomes mind becomes consciousness, and as it becomes consciousness, becomes responsibility, etc., etc.

Tobias: But that model seems to have broken down if one examines Neanderthal endocasts, for example, which had a luxury; an excessive neocortex in terms of cubic centimeters: 1600-1700 cc's as opposed to more modest and more thoughtful Cro-Magnon at 1200 cc's, just to take an example. So apparently it's not the number of cc's. After a certain threshold we cross the cerebral rubicon and seem to need no more than that. It appears that for the last 20-30 thousand years, cerebrally, no more grey matter has been accumulated but something else has been accumulating.

Soleri: I have thought about that. The brain's redundancy. Well, I tend to believe that the instrument has to come before the performance. Perhaps an instrument was developed, and the performers weren't ready. So the instrument was discarded; or set aside, or placed in a holding pattern. In addition, you need redundancy to insure survival. The earlier cortex might have come before its time.

Tobias: Interesting.

Soleri: Not a full explanation . . .

THE CITY AND THE CABIN

Tobias: I have this image of Chicago in my mind, and of Saul Bellow in Chicago and wind, and of the Chicago Art Institute and all of those wonderful parts of Chicago. And then I think of the south side of Chicago. You said that the increasing complexity might enable us on a spiritual and biological level to grasp more essentially our surroundings, which I interpret as being not confined to the human surroundings. From the heart of Chicago, twenty stories up, with six inch separations between myself and my neighbor, what potential do I have for enhancing the spiritual complexity with the outside environment? How do you perceive that division which has reigned for all time between ourselves and our surroundings?

Soleri: Probably, in order to have a discourse going on, we should say that this is one occasion and that's another occasion. So let's compare them. On one side, we are living 20 stories up in Chicago and on the other side we have the person in the forest living in the log cabin. I would propose that of the two realities the reality of Chicago is not only more pregnant, but it's more productive in spirit or mind. Not necessarily because the individual living in that apartment is more intelligent than the person living in the cabin, but because the person living in the apartment presumes a network of things which is per se a premonition of the appearance of a new kind of animal. It might well be that the person in the apartment might not be very intelligent, but the neighbor might be, more complex, let's say. But for me the person living in the cabin is handicapped because fundamentally the social and cultural foundations are very primitive and they might also be nonexistent. I have to assume for this comparison, that the person in the cabin is not living there while being plugged to Chicago, but rather is there because Chicago doesn't exist. So since Chicago doesn't exist, nothing exists of the technology that might help this person. He doesn't have books, because books were not invented. He doesn't have language, because maybe language has not been invented if he was isolated. He just has ways of building a fire and contemplating it.

Tobias: And yet, the predominate proportion of our species has been and is agricultural, living out there in a cabin (if they're lucky), living in a very frail concatenation of community farms.

Soleri: Yes, but they are tied to larger communities which in turn are tied to still a larger community. It's not the history of the cabin; it's the history of the cabin which is plugged to the community, which is plugged to the village, which is plugged to the town which is plugged to the city. So it's nice, as a Chicagoan, to get away from it because I've the means of getting away from the drudgery of it, but carry with me the advantages of it. So I have the radio connection, the television connection, the book connection, the telephone connection, the watch, the shirt, glasses, and I have all those things that have been given to me, not by the cabin phenomenon; it's a phenomenon which has the cabin phenomenon within it.

Gosney: Interconnected urban knots.

Soleri: Yes, because that's where the advantage of cooperation and exchange and energy and so on allows me to be critical of Chicago.

Brown: Could we switch that argument to the catch phrase "the global village." Is there not possibly a process whereby through again the technological process of communication and so on, the global village could gain in complexity within the cabin kind of syndrome we're talking about? Is there a paradox there in some ways that even in your cabin you could have access to all kinds of . . .

Soleri: Yes, if I was born in a cabin with all those privileges, books, etc., etc., then I probably would say, that's it. We've got it, why don't we carry on. If I was born in a cabin and then I migrated to the city, a good city, then I might change my mind and say "The cabin is good, I like to go back to it now and then, but boy, the city is where things are happening!" And if I was born in a city and had no notion about what the cabin is, I would probably say, well what's that little thing there, it even scares me. You take a boy from the central city, who has never been in a forest and he's scared to death really. It's a scary proposition. That doesn't mean he might not enjoy it. But at the same time it's a dreadful notion because it's a foreign notion. But I think that once you get used to all those experiences then it might well be you find that the cabin is a good vacation place, but a city's a good living place.

Brown: So what you seem to be saying is that it's the direct contact between individual humans that makes the city the place it is. It's not the access to information so much.

Soleri: Yes, I think even though we never talk about that, I think that's very fundamental if we're going to end up with brains plugged into computers, end up with something which might be very frightening. If you're able to plug the computer into human beings then we might end up with something far more exciting because we have all the senses there helping us to really enjoy what the computer is able to tell us. But if the computer is only going to be able to tell us something which has become another computer then there is not very much pleasure in it.

Tobias: But isn't that precisely the risk involved in the city conception of the universe?

Soleri: Well, that's why the city has to be altered. In fact, it's that which allows me to be very close to the country without losing the assets of the city. Which is really what the arcology notion is.

Tobias: What is that component of the country for you? You've referred to a spiritual and biological complexity which I assume refers to the outside, beyond the city.

Soleri: An essential component for reality to me is a sensorial reality, a reality which comes from that which I am the offspring of. And I need to go back to it.

Roxburgh: Which is kind of an inherent property, like gravity.

Soleri: That's why again, speaking of absolutes I am very intransigent and say well we have to get rid of it. How do we get rid of it and find the solution?

Tobias: Gravity?

Soleri: Gravity as one of the rules, as you were saying, that are obstacles to the spirit. But then I find out that gravity is also the maker of the spirit because without gravity we would not be here and talking about spirit.

Roxburgh: Right.

Soleri: So, if it's an instrument or a medium it is *intrinsically* unable to become pure spirit, but it's the only way by which spirit can be implemented.

COMPLEXITY AND THE URBAN EFFECT

Tobias: What is your principal motivation for pursuing the city of the future? Is it based upon your awareness of population problems or population reorganization needs or is it really your quest for a greater complexity, representative of the human spirit?

Soleri: The need of the urban effect is not generated by the constriction of space, or the fact that the planet is too small or that we are too many. That's only just happened recently, and how we have those *responsibilities*. It's positive by itself. It doesn't need justification, by the limitations the Earth is presenting us with.

Tobias: What are the justifications?

Soleri: The justification is that complexity, is, per se, divinity.

Gosney: Greater complexity, greater perfection.

Soleri: Yes.

Gosney: It seems these limitations on the planet are going to bring people to some of the conclusions that arcology is suggesting.

Soleri: Yes, I believe so, I wish I could make it a little more forceful, but I think that's it.

Tobias: Complexity as that totality of interactions between human beings in greater and greater density.

Soleri: Yes, but in order to make sense out of it, I have to start at the very beginning. If there are two particles which are hitting each other but they are somehow so ermetic that they cannot communicate. They just hit each other, then I say it's a complicated affair, but it doesn't trigger something. Then a little window opens in one and another little window opens in the other and they see each other, and talk to each other. Then you have a triggering of a complex process. So already, there is an intimation that something more is there than just hitting each other.

Tobias: But they level off. Every species levels off, every community levels off in the amount of absorption of energy that it can receive. It can't keep increasing indefinitely.

Soleri: But evolution contradicts that in a way, since at the beginning there were those few little agitations that were not much more than physical, mineral agitations – and now we're people agitating in tremendous ways. And agitation becomes animation at a certain point. I think, again, because of the summing up of so many events going on in the same amount of matter that at a certain point the "shaking" becomes mind.

GAIA

Brown: Do you see no limits to the growth of mind? As Michael was saying, populations limit themselves. There must be an energy limit, as it were, of the biosphere that the Earth can support – in terms of mind. Or do you just see this as a growth that goes on and on? Maybe the Earth is not an aggregation of individual organisms like us; but as the Gaia hypothesis proposes, the Earth is the organism and as many cells as that organism needs to operate is the optimum.

Soleri: Okay, well, that's where the Christian background, the notion of equity comes in, and says that's fine, but it's not sufficient. I cannot accept that you, me and this olive tree are our only means to something. I have to find a model which tells me that ultimately the tree, you and I are ends in themselves. How do I find an answer that satisfies this?

Roxburgh: It seems that the perception of an environment or of oneself is fairly important, an important process. The process of perception itself is really important somehow.

Soleri: That's right.

Roxburgh: And the intensity of perceiving what we're doing seems very valid – in the moment.

Gosney: Perhaps a possible definition of complexity is our ability to enfold the encompassing systems. First we know ourselves – self-reflection of consciousness. Then we know our society, our species. Finally, the Gaia hypothesis is getting to each human being as conscious, the planet being fully conscious, being conscious of this whole system. The next step seems to go into even more cosmic perception. But it all is just that act of perceiving.

Soleri: Somehow the Judeo-Christian model keeps insisting on the nature of the individual being an end in itself, the importance of the person. I cannot dismiss that, so I have to find a way by which Gaia and Christianity can come together.

Tobias: To go back: in terms of energy leveling off, remember that the Sun's energy is only 1% efficient in terms of the total surface of the planet, the rest is dissipated. It can't be effectively utilized. Which initially suggests to me that there is an optimal limit to what degree of interaction can be absorbed and what cannot. From another perspective, we could look at the sexual revolution a billion years ago with the eukaryiots and the feverish dance that was put forth throughout the marine environment of the planet, the voracious, redundant Fantasia to the fortieth power; it was a remarkable amount of energy which succeeded over a half a billion years in multiplying and qualifying to simpler hypertropic levels. This continues on the inside – but we do not as organism per se spend our days in such feverish rhapsody. Maybe some of us would like to; but here as complexity succeeds complexity, topples complexity, it tends to go at *simpler* constructions in order to manifest itself. This is not to say that we abandon the bottom rung. We don't, the bacteria stay with us, they're in the gut of most organisms. But *we* become, in many ways, simpler, more elegant.

Soleri: I have a problem there with Gaia and all the people who say that the real thing is the genes or the elementary particles, and so on because they are the only things that perpetuate themselves, some even live eternally, cannot be destroyed, etc. Those elements, it seems to me, might be the bricks by which we build a structure which is more than the sum of the bricks. So I'm not buying the fact that we are only carriers of something, that we are disposable. The complexity is not how complicated or complex is the brick, but how the castle, or the building is complex, multidimensional, and so on.

You mentioned the one percent (of the Sun's energy) falling on the Earth. That seems to be confirming the fact that nature, from the very beginning tends to be enormously disinterested in efficiency, in many ways . . . "Who cares?" The sun and 99% of the energy somehow – at least from an anthropocentric point of view – is totally wasted. (Probably it's not wasted.) But, we could say that since in the solar system it seems that only on this planet is complexity building up in dramatic ways, nature has to transform itself to respond to this nascent birth and this exponential development. We *are* nature, and we have to find ways of containing that wastefulness and make it into usefulness.

Tobias: As a model, as an image by which to propel ourselves into the future, you say that nature is extravagant to the extent that she's not concerned, particularly, with usefulness; and it becomes incumbent upon our species, in all our creativity and spiritual prowess, to become useful.

Soleri: Which is the notion of genetically, fundamentally, putting order, where a lesser kind of order existed.

Tobias: Do you think that this is so; that there is a rage for order that is implicit in all people?

Soleri: A what? A rage?

Tobias: A *rage*. Auden's expression, he wrote it in a poem, a lovely work, but something I've always thought about as being not necessarily universal. And one could not simply delve into the anthropological realm to seize upon little expressions of disorder.

Soleri: It depends how we interpret the reality of the organism. Suppose you have an anarchic tendency, you have a tendency of total disorder, you live like a pig, let's say. Well, to you, it's order of a sublime nature. No matter how piggishly you live your life, you are order of an incredible complexity, because when you become disorderly, even slightly disorderly, as a physiological creature, you're in trouble. So you personify order and hierarchy, and, again, cooperation and interdependence, and all those things. And, in a way, if you like to betray that – that order gives you a right to do this. But it's a betrayal.

Tobias: A man I knew named Carlos, a Jesuit philosopher from Madrid, always used to say, when he really got into a corner: "Ah, between my accent and my penmanship, there is no hope." (laughter)

Soleri: That's a good slogan.

Tobias: He was marvelous. He would write on the blackboard and it was truly Phoenician – illegible all the way. When we talk about complexity – in the city, as being its proper place – I keep thinking of people like Thoreau and Lucretious, whose minds brim with contradiction and exquisite civilized and civilizing complexity, from their cabins and their farms – not tied to any central market place – to any Chicago.

Gosney: They were products of their very upbringings. They were products of the culture created by the urban. . .

Soleri: Yes, that's number one. Number two is that they are – we are – urban effects ourselves. So, it's very admissable and often very admirable that we can be so self-contained as not to need the envelopment of the urban effect.

Tobias: How admirable, how enviable, how important is that?

Soleri: I think it's important, but it's also self-limiting as one can do on one's own up to a point, and then it falters. By nature we are segregated urban effects and segregation is ultimately always limiting.

Tobias: What do we need?

Soleri: We need other urban effects to touch and to be in touch with. We need to work around institutions which urban effects are devising, such as the theater, the church. . .

Tobias: Does everyone need the theater?

Soleri: No, not everyone needs the theater, but take the theater away, take everything that relates to the relationships among people away and you have very little left.

Tobias: Every one of us here – by the Willy Brandt report findings – consumes 1042 times more, per capita, than every Nepalese. This degree of redundancy which we demonstrate seems to me to be inefficient. It is an unnecessary replication of the Sun's own inefficiency as we perceive it. When we speak of necessity in our life, (things that we as human beings need) I'm reminded of the doorknob and of the bed which have not changed – size-wise – in thousands of years. Apparently, it's a practical and efficient size that relates to the human being in an intimate way. So I'm asking you, is there not a whole range of intimate requirements which need no more complexity?

Soleri: Yes. You see, if you demonstrate to me that you could have the Lucretiouses and the Thoreaus and the Einsteins and the Beethovens and the Stravinskys, etc., without this association and this cooperation, without this social nexus and this cultural nexus, then I would say, "Maybe the cabin is the answer!" But you wouldn't be able to demonstrate this. Thus, the fact that a Western mode of living consumes a hundred-whatever-thousands more than other modes of living doesn't as yet tell me that this Western mode is more wasteful. I have to see the results. And again, I might be very parochial, but if I have to choose between what the West has given to me and what the Nepalese has given to me – has given to mankind – I would have to choose the West. The reason is because the West has given me Beethoven and a few of those other guys.

Tobias: So, for those momentary zeniths . . .

Soleri: Ah! Those are not momentary . . . those are everlasting! If there is anything that is everlasting and an absolute in many ways, it is the spirit exuding from those zeniths.

Radin: I'm beginning to feel defeated. That is a classic . . .

Tobias: And the trade-off?

Soleri: Economy . . . I think the epitome of economy is not to be found in what we call the economic world. It is to be found in the aesthetic world, where a very tiny amount of energy, a very tiny amount of material, does very powerful things. By the way, did you . . . see *Tristan and Isolde* last night on Channel 8? There was this black singer; did you see it?

Roxburgh: No, I didn't see it.

Soleri: This incredible singer played Isolde. She was marvelous!

Roxburgh: Was it Price?

Soleri: No, it was a younger woman. I have seen her before, but in this part she was . . . ahhh! That's when you feel that it's all worth it; it's worthy being alive.

Roxburgh: Yes, there's a continuity in music, I think. To me, music seems really tangible – like a thread throughout civilization. Some of the earlier folk music turn up today, in all sorts of modern music.

Tobias: What is your greater urgency at Arcosanti? Is it the aesthetic pedestal or is it a form of rational survival? I grant you that it's both.

Soleri: I always believe that the two are working together. But, just for thinking, you say well, if I cannot do the most, let's see if I can do the least and the least would be to offer alternatives that might be more useful . . . to life. And one of the usefulnesses is to be less encroaching on the other kinds of life which are around.

Tobias: So the non-intervention, less intervention does play an important role at Arcosanti.

Soleri: Lesser intervention.

Gosney: More intelligent intervention.

Soleri: Not because I don't believe in intervention, because anything that exists, any organism is an intervention in something which existed before the organism. But because I think we are still quite unable to really come out with a realistic response to the historical and contextual moment. If I were superintelligent, superwise, and superloving, I would say, well, I feel that I can intervene much more substantially. But since I am not any of those, I have to be prudent.

ARCOSANTI

Tobias: What are some of the basic tenets of Arcosanti which are "lesser interventionist"?

Soleri: The size itself: the idea that by reducing the bulk of something you might be able to see that the sum of everything including the ecological system, is better served. The fact that since we are not pure spirit, we depend constantly on gravity laws, thermodynamic laws, time and space constraints; therefore, again the notion is that the complex is utterly connected with size, which is miniaturization. The fact that perhaps we can do better with existing energies, mainly solar energy in this case, to help us to take care of our needs. The fact that in some areas of the globe, water is scarce, so becomes precious, so we should try to pattern our habitat in ways which are saving and make better use of water. And then the one which I think is in many ways the resume of all of the other ones, that *things tend to segregate themselves,* in many ways because they had to, because they are not good at *not* being segregated.

Tobias: That last point, again?

Soleri: I stand for segregation in the sense that I had to isolate my flesh from your flesh and from what surrounds me in order to exist; this segregation gives me not only autonomy but makes me real. If I take a knife and split myself in little slices and say now I am connected, well, sure, that which composed my body is more directly in touch with other bodies, but I no longer exist. So I have to segregate those elements in a way to exist. But then if that gives me justification for applying this notion of limiting and separating and dividing to other contexts, I find out I'm undoing what I'm probably here to do – which is to synthesize and get things together instead of separating and splitting things apart. So as an autonomous creature, I need to exist with other autonomous creatures. Constant connection – that's the urban effect.

Tobias: Is there some optimal density of population that would be standard throughout the world?

Soleri: I think that probably doesn't work in space-time parameters and doesn't work historically. Each context has its own answer. On the planet now there are all sorts of optimal conditions for different kinds of contexts.

Tobias: Are those conditions determined by the environmental constraints? Or are there other determinants?

Soleri: Only in part environmental – then also social, political, economic, technological, psychological, religious, philosophical.

Tobias: So they all play an important role?

Soleri: Yes.

DEEP ECOLOGY

Gosney: I'd like to suggest that we address the whole idea of deep ecology. Paolo, were you able to read Arne Naess' paper, which is part of this book?

Soleri: Yes, the one you sent me some time ago? Yes, I read it; but my memory is a little vague.

Gosney: Are you familiar with . . .

Tobias: He was asked to be part of the group, the New Natural Philosophy . . .

Soleri: Yes, Joseph Meeker asked me, two or three years ago, in a meeting with Arne Naess, if I wanted to be part of this and I said "Yes, but as you know I'm . . ."

Gosney: Entrenched in your own work . . .

Soleri: "I might be your worst enemy, I don't know."

Gosney: Well, I'd like to know your feelings about this idea of deep ecology; I think it's a concept that's still forming, but as a newer expression of the ecology movement. As I understand it, the basic suggestion is that we begin thinking more *as nature*, man *as* nature rather than man *and* nature, which is more the shallow ecology, as they say. How would you compare what you're doing at Arcosanti, an embryonic "urban laboratory" as you've called it . . . how does your work there relate to the deep ecology movement?

Soleri: I'm in total agreement, at least intellectually, in saying that it's not "man and nature," but it's "man as one of the manifestations of nature." And then I would immediately have to add that man has added a demonic dimension to the natural process; we have to keep that in mind because that alters the nature of Nature. The alteration might be very localized, maybe in one spot and then billions of spots . . .

But if we consider that as evil or if we dismiss that, then I think it becomes the old notion of Nature which is the Garden of Eden and what was before the Fall, before the origin of sin and so on . . . at that point I would say I'm not with you any longer. Consciousness has triggered something that in a way explains why now we have man *and* nature, because the appearance of consciousness somehow is demonic or shall I say Faustian.

By our appearance, reality has changed its own nature. It's not nature plus man, but the nature of man which redefines nature. As perhaps the prophets of some religions have defined, including the Taoists.

Tobias: Is there some explanation, other than the "demonic" one that would explain the nuclear generation and the advent of the bomb?

Soleri: Yes, perhaps, the demonic is too demonic. What I mean is that once there is the development of something reflecting on something else, and developing a consciousness of this reflection, we have the beginning of the notion of good and evil – not only the notion, the *beginning* of good and evil.

Tobias: So, it's beginning now.

Soleri: It began as soon as the cortex began . . .

Gosney: Reflective consciousness . . .

Soleri: "I do not do, I calculate and then I do." Calculation is positive or negative.

THE BOMB

Tobias: You spoke of quantity as somehow indicative of a higher and necessary complexity. I think of the dinosaurs' size and its ultimate extinction and the obvious analogies to the A-bomb and nuclear weaponry. How do we redefine that advent so as to remain optimistic in this whole progression to which you've been referring? Your own descriptions suggest a necessary logic for the development of the A-bomb, and there are those around us who govern us, who insist on the logic. But according to your scheme, it seems to me that that is such an obviously suicidal course for the survival of all of these Beethovens and human interactions that we're seeking – something went awry.

Soleri: Yes, but I keep telling people who work with me to perhaps a nauseating degree, that inequity is ingrained in reality and is particularly ingrained in human reality. So the atomic notion, in terms of the use we are planning to make, is evidently clear evidence of the inequity that is built into ourselves. For the case of atomic war it is an inequity which develops out of fear and bigotry, again, intolerance. The notion that part of the human species maybe knows what goodness is but says "No, I'm not going to 'use it,' I'm going to be something else."

Tobias: How do we reorient perception? Given that it accumulates from the various components, the religious, the sociological, the psychological, the ecological, the aesthetic: they combine to create this dependency, this paranoia . . .

Soleri: That's where the responsibility of the person comes in, and the importance of the person. I have to see if I am very coherent when I decide that it's better for "you to be dead than for me to be red," for instance.

Tobias: This takes us full circle. I see you here contemplatively under your tree and I think that the individual to whom you now refer is the ultimate salvation in terms of being responsible.

Soleri: But the individual becomes a universal person, it doesn't remain just a little creature under a tree; it becomes universal enough to understand that even though I have my preferences, those preferences have to be within the boundaries of the preferences of the species. And the preference of the species is not self-destruction.

Tobias: So what separates a Ninth Symphony from a Mein Kamph – in terms of that will to universalization?

Roxburgh: That's right, because there is sort of a reflection say in the Ninth Symphony with Beethoven being driven to communicate; and maybe somebody like Oppenheimer is, in some sense, an artist too.

Tobias: Absolutely.

Roxburgh: It seems really important as far as deep ecology is concerned, that we choose our materials and really look at the motives behind and the implications of our art or craft or passion. In a way, I think Oppenheimer was commissioned – the population thought it was a righteous cause . . .

Tobias: The euphoria in the U.S. following the A-bomb in Hiroshima.

Roxburgh: Yes; at a certain point I don't think there was any need for Oppenheimer to go on. He just felt compelled to finish the project. It just seemed like that same obsession that would charge a Pollack or Beethoven or Stravinsky – it's sort of inherent in anyone that taps it. It seems like we all have to be real careful at this point in what materials we use.

Tobias: Yes. But it was not just Oppenheimer. There was a commission that Roosevelt, Truman put together that included the president of Harvard University, presidents of banks, humanists . . .

Roxburgh: The population was pretty well behind it.

Tobias: Completely.

Brown: That's like any organism in a sense. You could say that each of us is a conglomeration of individual organisms; their consensus is what we are. Just the same in a social setting, that there are clusters of organisms and the consensus of that group is what these organisms are. What we have to take account of is trying to define the moral and ethical basis on which those clusters or groups are based, which is sort of like producing the conscience of the new society.

CHICAGO / ARCOSANTI

Tobias: It's precisely that point where I both take issue with and subscribe to the whole notion of Arcosanti. What I mean by that is when you too easily use the analogy of a Chicago, I get nervous. Because then I begin to feel that the individual is compromised and that there will be no hope amid such a pervasive density – an individual autonomy rising to universal morality. However, when the basis for that Chicago, from its inception is made clear and made persuasive, then I'm ecstatic because I sense great hope.

Soleri: We have to keep in mind that Chicago, like any other city, is not only embryonic, but very pathological in many ways; it has pathologies scattered all over inside of it. So even though you admire the sick dog, you wish the dog wasn't sick. But to move from the sick dog to the healthy dog is going to take a tremendous amount of work.

Tobias: Yes.

Soleri: The fundamental notion is absolutely right; but what we're doing with it needs a tremendous amount of work.

Tobias: Can you graft your lovely structure out in Mayer, Arizona to a Chicago, or does it have to start from a beginning?

Soleri: That's like comparing a virus to afish; Chicago is far more complex than Arcosanti. It also happens to be more complicated.

Tobias: But how can the ideas of Arcosanti become applicable to Chicago?

Soleri: That's where the laboratory notion comes in. You start in a simplified way. You are trying to take the village idea and make it into more of an urban condition. If you are able to trigger that kind of transcendence, or a little quantum jump, then you can go to Chicago and say, "What about trying to apply these notions and see what happens."

Brown: Arcosanti can become this little virus and begin to penetrate other cells and start reproducing itself . . .

Tobias: William Burroughs has used that idea in *A Naked Lunch*. He speaks of "the word virus" – a single word with the proportions of God, the gravity of God – which, within 24 hours has infected every speaker on Earth. We all know, for example, how fast a joke, or pheremone gets passed around. Nervous impulses in Homo Sapiens travel 224 m.p.h.

Gosney: A concrete example of what Arcosanti is exemplifying might be integrated into Chicago . . . I'm sure somewhere in Chicago there is a new condominium/shopping complex being built right now. Just take one of those, the energy that's going into that and apply a few of the simple principles that are being shown at Arcosanti, for instance, design a apse shape into it and use the position of the structure in relation to the sun as a heat source.

Soleri: Yes, *[several minutes lost on tape]*

Tobias: But I'm not at all disposed to this "virus technology." I feel it is imperative to investigate pre-virus humanity, before it has been usurped. There are still some tribes throughout the Transhimalaya or the Amazon, for example, where there have not been many migrations over the last 5000 years. And yet, today as we probe into cultures that reside there we find remarkable insights in their agricultural techniques, marvelous handicrafts and a whole host of other human, specifically human activities which are incomparable. Amazon Indians can lavish on their verb tenses greater expressiveness – call it complexity – than Plato, Truman, Mozart. A human being with sensitivity and artistry is producing something which has integrity about it in the heart of the Amazon. The point there is, the old Socratic argument that we've been touching on throughout our conversation: the person seated under the olive tree – do we really know anything about him? Is his gift to fellow people as relevant, as fiery, as vital, as your poet in the upper east 60s of Manhattan, with the constant onrush of sensation . . . Or what of the Tibetan poet vs. the atomic physicist?

Soleri: I think that if we contrapose that group there to this group there, we should not just talk about the poetry of both of them – we should talk about what they have been "producing." As soon as we look at the bulk of "production," I would say that it's very interesting that

PHOTO: Robert Radin

Paolo Soleri at Cosanti

somebody in Tibet has the notion that somehow talking to us in the same vein as the particle physicist – I would say that the particle science is going far beyond the notions which are similar to what happens in Tibet. As soon as I begin to look at the power and the bulk and the number of things that Western science has been giving us, I *cannot* see anything comparable. And music is the same.

Tobias: How much of that riotous science are we using? Let's be specific about Arcosanti. You have at your disposal hundreds of alloys, hundreds of possibilities for construction material, but you use a very basic, simple one.

Soleri: But when I am through with casting concrete, I go home and open a book and read about the Gaia hypothesis, I read about Prigogine and think about that, and Einstein and Fermi – it's an incredible, gigantic manifestation of mind . . .

Tobias: You don't think it was like that 5000 years ago?

Soleri: No, but then the detail. Picking up that blade of grass and investigating the universe which is in that blade of grass and beginning to develop notions about the nature of it and how it's there and how it develops, how it was conceived, how I was born . . . those things are enormously exciting.

Brown: So you see that as more powerful though, than say Blake saying, the world is in a grain of sand. He's saying the truth, but what you seem to be stressing is the investigation of why that's true.

TECHNOLOGY AND SURVIVAL

Soleri: The reason is that ultimately I believe that anything that becomes known and any technology that we develop is for the sake of the spirit, because it is that which is going to give us the means to manipulate and transform reality. So the power that science is giving us, which we are mis-using most of the time, is still a gift of such dimension that we should be constantly astonished.

Brown: But the same intuition comes from a poet . . . you are saying the intuition coming from science is more powerful?

Soleri: It's not the same intuition. One is the intuition of some kind of second or third sight, but the other gives me the instrumentation to make use of that intuition. I am a believer in technology, a profound believer.

Tobias: To believe in technology is fine, in my opinion, but when confronted with technology, how many human beings can utilize it knowingly? We go to the supermarket and we buy labeled plants. What do we really do with our technology as individuals?

Soleri: I think the fundamental thing there, is that we are overwhelmed for the moment by what we invent, and we have to get used to it. We are not geared up to it yet.

Gosney: To make truly intelligent use of it.

Tobias: But you do believe we will get used to it.

Soleri: Well, when I see a child that goes knowingly to the computer, I say "Whaaa. . .?" (laughter)

Roxburgh: We really are at a point right now where, with such a rate of technological progress, it seems that man is going to have to adapt and find what's really useful . . .

Gosney: Come to grips with it.

Soleri: Yes, and my notion is that the more we get into this power of technology, the more we should become wiser and more knowledgeable and compassionate. That is why I always need to connect that with the environmental. This is the reason I believe habitat is so important, because it gives us the pedestal for this environmental learning process that is almost at the opposite end of the learning process which is given to us by science and technology.

Radin: But does man as he gains power gain ethics?

Soleri: No, they don't go together it seems. In fact, I tend to believe that in general power corrupts, but not necessarily.

Tobias: And is power technology?

Soleri: Yes, power of transforming . . . since the knowledge of the biological system is the knowledge of bio-technology of transformation. Since I admire very much the bio-technology of evolution, I cannot be persuaded that now that we have a part of biological technology, we are dealing with evil. We are just dealing with a new facet of technology which is too young to be wisely used.

Tobias: Do you feel we will find a peaceful resolution to the present war that's going on, using technologies, between the East and West?

Soleri: I think peace is far off, but there are a number of responses which are not necessarily catastrophic that might be in the offing. I don't believe in the distinction between that which made us what we are and that which is going to help us become something else . . . and that which is going to help us is a new kind of technology. But it's still technology.

Tobias: And what made us was technology?

Soleri: Yes. The ability of something to manipulate and transform something else, for the sake of some goals or some performance.

Tobias: Yes, I understand in Salt Lake City, Utah, there is a company dealing now in condominiums for post nuclear circumstances.

Soleri: I saw something about that, yes . . .

Gosney: We were talking on the way down about the incredible power of the individual's survival instinct, and hoping, I guess, that the species can manifest that same kind of power. I noticed you mentioned Prigogine; I guess this is a slightly esoteric divergence from the critical issue of nuclear war, but it does seem like the whole human species, looking at it as a system, is at a critical point right now, having reached in many ways a critical mass. The Prigogine theory of biological complexification is an interesting analogy: a system, before it rises to a new level of complexity, seems to fragment and start to fall apart, looks like it's going to dissolve . . .

Tobias: Dissipate.

Gosney: Hopefully . . .

Soleri: We're going to make it.

Gosney: Through the media and increasing interconnectedness of the planet's natural and technological resources, we're going to make the right decisions, work on the new order.

Soleri: The question is to see if the deep ecology movement is going to be helpful.

Pen and Ink by Nathan Weedmark